▶ Brain, Mind and Internet

DOI: 10.1057/9781137460950.0001

Other Palgrave Pivot titles

Shane McCorristine: William Corder and the Red Barn Murder: Journeys of the Criminal Body

Catherine Blair: Securing Pension Provision: The Challenge of Reforming the Age of Entitlement

Zarlasht Muhammad Razeq: UNDP's Engagement with the Private Sector, 1994–2011

James Martin: Drugs On the Dark Net: How Cryptomarkets Are Transforming the Global Trade in Illicit Drugs

Shin Yamashiro: American Sea Literature: Seascapes, Beach Narratives, and Underwater Explorations

Sudershan Goel, Barbara A. Sims, and Ravi Sodhi: Domestic Violence Laws in the United States and India: A Systematic Comparison of Backgrounds and Implications

Gregory Sandstrom: Human Extension: An Alternative to Evolutionism, Creationism and Intelligent Design

Kirsten Harley and Gary Wickham: Australian Sociology: Fragility, Survival, Rivalry

Eugene Halton: From the Axial Age to the Moral Revolution: John Stuart-Glennie, Karl Jaspers, and a New Understanding of the Idea

Joseph Kupfer: Meta-Narrative in the Movies: Tell Me a Story

Sami Pihlström: Taking Evil Seriously

Ben La Farge: The Logic of Wish and Fear: New Perspectives on Genres of Western Fiction

Samuel Taylor-Alexander: On Face Transplantation: Life and Ethics in Experimental Biomedicine

Graham Oppy: Reinventing Philosophy of Religion: An Opinionated Introduction

Ian I. Mitroff and Can M. Alpaslan: The Crisis-Prone Society: A Brief Guide to Managing the Beliefs That Drive Risk in Business

Takis S. Pappas: Populism and Crisis Politics in Greece

G. Douglas Atkins: T.S. Eliot and the Fulfillment of Christian Poetics

Guri Tyldum and Lisa G. Johnston (editors): Applying Respondent Driven Sampling to Migrant Populations: Lessons from the Field

Shoon Murray: The Terror Authorization: The History and Politics of the 2001 AUMF

Irene Zempi and Neil Chakraborti: Islamophobia, Victimisation and the Veil

palgrave▸pivot

Brain, Mind and Internet: A Deep History and Future

David J. Staley
Ohio State University, USA

DOI: 10.1057/9781137460950.0001

First published 2014 by
PALGRAVE MACMILLAN

Palgrave Macmillan in the UK is an imprint of Macmillan Publishers Limited, registered in England, company number 785998, of Houndmills, Basingstoke, Hampshire RG21 6XS.

Palgrave Macmillan in the US is a division of St Martin's Press LLC, 175 Fifth Avenue, New York, NY 10010.

Palgrave Macmillan is the global academic imprint of the above companies and has companies and representatives throughout the world.

Palgrave® and Macmillan® are registered trademarks in the United States, the United Kingdom, Europe and other countries

ISBN: 978-1-137-46096-7 EPUB
ISBN: 978-1-137-46095-0 PDF
ISBN: 978-1-137-46094-3 Hardback

www.palgrave.com/pivot

DOI: 10.1057/9781137460950

Contents

Preface

However you may feel about its politics, Garry Trudeau's *Doonesbury* is particularly good at identifying broad social trends in our culture. One of my favorite strips takes place in a college lecture hall. A student sits in his chair furiously typing away at his laptop, obviously distracted from the lecture. His device pings him: a fellow student warns him 'Head's up dude – professor just asked you a question.' None of his friends seems to know what the question is, since no one in the room apparently is paying attention to the professor. One of the student's electronic classmates overhears the question and chats back 'name four major greenhouse gasses.' The student pings his friend 'stall her while I Google the answer.' From the back of the auditorium, we hear 'Professor, we couldn't hear the question back here, could you repeat it?' 'I asked Mr. Harris to name four major greenhouse gasses,' replies the teacher, after which comes the immediate reply from Mr. Harris 'Water vapor, CO_2, ozone and methane.' The professor concedes 'uh ... right.' Triumphantly indifferent, our student chats back to his friend 'If this keeps up, I'll never get through my email.'

I have shown this cartoon to a number of teachers over the last few years. They immediately light upon the behavior of the students, and decry the use of laptops in their classes. The students are distracted from the class, are having their attention drawn away from the lecture, indeed are not even engaged, but are rather chatting with friends or emailing or websurfing. When pressed, teachers express concerns over classroom management and control in

DOI: 10.1057/9781137460950.0002

such an electronic setting. Not only are the students distracted from the lecture, this appears to be a coordinated distraction, as the students are engaged in a subterranean conversation extraneous to the formal class. The students come across as wily and duplicitous, conning the professor with their clever use of technology.

But upon deeper examination, many teachers also see an issue with the professor's behavior, or at least with the pedagogical architecture of the class. The professor has asked a relatively simple question, one that can be easily looked up. The underlying pedagogical assumption as expressed in the behavior of the professor is that the student, having diligently read the material the night before or having been attentive to the lecture, should have such information ready to recall at the professor's insistence. This is, of course, a standard way to think about education: information and knowledge is deposited in student minds, ready to be recalled upon demand. (It is the underlying logic of standardized testing.) The student clearly does not have this information at ready recall, in his memory, at any rate. A quick check of Google, of course, yields the answer as quickly as if it were embossed upon his memory. Teachers understand the implications: if such answers, if such information is so readily available on the Internet and if students have easy access to that Internet through a laptop or some other portable device, perhaps the 'deposit' model of teaching and educational assessment needs to be reexamined.

The present moment might be described as an era of 'just-in-time knowledge.' With smaller and smaller devices capable of accessing a cloud-based information infrastructure, it becomes increasingly unnecessary to retain some information in our physical, biological memory alone. As more of our symbolic knowledge, our cultural storehouse, can be migrated from libraries, archives, and museums and uploaded to the Internet, and since we can easily access that knowledge wherever we might be, we need only query this 'external memory' when called upon and on-demand. In the same way hand-held calculators freed us from having to memorize times tables, a mobile 'library in our pocket' frees us from having to retain some information in our biological cognitive apparatus, thereby expanding our cognitive capacity.

Observers remark that we live in the era of 'cloud computing,' but this essay makes the case that we have been surrounded by an informational and symbolic 'cloud' for much of recorded history. Human have, for millennia, had an *intimate* relationship with tools that expand cognitive

DOI: 10.1057/9781137460950.0002

capacity. When determining how cognition occurs, the brain and cognitive tools *together* engage in cognition. Accessing the Internet is but the latest iteration of a cognitive act whose origins lie in the prehistoric past.

Some observers find this system of just-in-time knowledge to be disquieting. John Brockman asked his *Edge* contributors 'Is the Internet Changing the Way You Think?'[1] Nick Carr has wondered 'Is the Internet Making Us Stupid?' by reconfiguring the brain's wiring system and making us all attention-deprived.[2] The present essay addresses these concerns by placing the brain–Internet interface within a broad historical context: that the Internet represents the next stage in a very long history of human cognition. One could, of course, interpret the Internet in the context of the medium-term scale of the 'electronics communication revolution.' This would situate the Internet within an historic context defined by television, radio, film, and, before all these, the telegraph. Had Marshall McLuhan lived to see its explosive growth, I am certain that he would have placed the Internet centrally within the electronic Global Village he identified in the 1960s. Many commentators and critics of the Internet situate it in relation to the Book; that is, they use the culture of print and the habits of mind it has enforced – McLuhan's Gutenberg Galaxy – as the context for understanding the historical meaning of the Internet. But I have been more influenced by Daniel Lord Smail, who asks historians to extend their notions of historical time deep into the Paleolithic.[3] Smail makes this temporal move in particular so he may use the history of the brain as a way to organize the narrative of such a 'deep history.' Since one of my goals in this essay is to understand the meaning of the Internet's effect on the brain, the best way to understand these changes is to situate our present moment at the appropriate scale of historical understanding.

Considering the brain–Internet interface within the context of 'deep history' provides a corrective to the distorting effects of 'Internet time.' Andrew Odlyzko identified Internet time as 'the perception that product development and consumer acceptance were now occurring in a fraction of the traditional time. Closely related to the concept of Internet time was the idea of 'first-mover advantage...' If indeed seven years of traditional product cycles were now compressed to one year, then anything might change in the blink of an eye.'[4] Even before Internet time, the culture of computing had enforced a sped-up notion of time; Gordon Moore's Law – the assertion that computational speed and processor powers double every 18 months – conditions us to think of technological change as

DOI: 10.1057/9781137460950.0002

quick and instantaneous, enforcing a time horizon of only a few months. I would like to interpret the meaning of the Internet by situating it not within the nanosecond scale of electronic culture, but the 'slow time' of the historical long term.

Following the approach of Fernand Braudel, I would like to situate 'Internet time' within the scale of the *longue duree*. Braudel divided historical time into three scales: the scale of events, where change occurs rapidly; the scale of broad cyclical changes, and a long-term scale of time he called the *longue duree*. 'Traditional history,' wrote Braudel, 'with its concern for the short time span, for the individual and the event, has long accustomed us to the headlong, dramatic, breathless rush of its narrative.'⁵ Braudel considered history from the scale of historical processes that were so slow or that extended over centuries that they were, in effect, stable structures that appeared to undergird the rapid movements of events. My particular interest in placing the Internet within a long-term historical scale is to draw our attention away from the dramatic, breathless narratives that 'Internet time' enforces on many commentators.

The human 'architecture of the mind' consists of both the biological brain and the cognitive tools we have developed to extend our minds. Humans have 'offloaded' cognition from the moment we started storing our thoughts in permanent symbolic form outside our bodies. Since the first Venus figures, body paint markings, and cave paintings, humans have been devising ways to create and store symbols in visible, external form outside of the biological brain, symbols that we can thereby access, exchange, interpret, and share. These external, materialized symbols serve as a cognitive prosthetic, extending our cognitive abilities beyond the limits of our biological brains. The brain conceives and constructs tools of cognition, such as art and writing and books and libraries. These cognitive tools, in turn, have reshaped the very mind that conceived them, a process that has spanned millennia. The Internet represents the next great extension of the 'external symbolic storage system' humans have developed since the beginnings of civilization. For all of the dramatic and disruptive change that the Internet surely represents, placing it in this long-term historical context renders this change more familiar, perhaps even less jarring.

Understanding that there has been a deep history of intimacy between humans and their cognitive tools provides us a framework for thinking

DOI: 10.1057/9781137460950.0002

about the possible futures of the brain–Internet interface, the future of the architecture of the mind. That deep history suggests that:

(1) The mind and its cognitive functions extend beyond the brain, beyond the human body. For millennia, human cognition has been both biological and technological. Indeed, our evolution has been determined as much by technological change as it has by genetic change. The subsequent development of the Internet, as well as tools by which we will access the Internet, will further extend the capacity of the human mind.

(2) The emergence of the Internet and its effects on our brains are neither a radical departure nor a cause for alarm, but part of an historical/evolutionary pattern whereby the brain begets tools that extend cognition that in turn, reconfigures the brains that beget them. That complex relationship between the brain and the cognitive tools it secretes will shape the future of the Internet.

(3) The future development of the Internet will be structured by the limits of the brain. Researchers studying how the brain learns to read note that, despite their wide variety, the world's writing systems nevertheless share common morphological features. All written scripts, whether alphabetic or idiographic, are built from a limited number of strokes, and occupy a similar bounded graphic space. The limits of the brain (which had to be refashioned in order to read and write) have shaped the structure of writing systems. There is reason to believe that the future development of the Internet will be similarly structured by the limits of the brain.

(4) External symbolic systems have historically been layered upon the pre-existing architecture of the mind. Writing did not displace speech, speech did not displace gesture. Each were symbolic systems that extended the architecture of the mind, with these earlier systems become vestigial to the new system. The Internet will not displace the literate/print-based mind; we will not completely lose our ability to read and concentrate, just as we have not lost our ability to communicate via gesture and body language, these ancient, but now vestigial, forms of cognition.

(5) The impulse to create a 'global brain' long predates the emergence of the Internet. The 'encyclopedic impulse,' the desire to collect all of the world's information and make it readily accessible, explains the development of universities and libraries and encyclopedia. The

DOI: 10.1057/9781137460950.0002

Internet as a form of electronic external memory is the expression of this encyclopedic impulse, an encyclopedia that is as once as vast as any library ever constructed and as assessable and portable as any book.

While the historic context of this essay goes back several thousand years, I do not intend to look forward thousands of years, but only about 25–50 years into the future. I wish to explore three possible scenarios or paths the future of the brain–Internet interface will travel.

Scenario 1: The internet responds to our 'queries,' it works when our brains activate it. The scenario represents the dream of H. G. Wells, who a century ago envisioned a 'World Brain,' by which he meant an encyclopedia created by the best minds in the world and made readily accessible to everyone. Wells imagined access to this Encyclopedia via books or microfilm, but with the Internet we now have the capacity to build the largest encyclopedia ever constructed, an encyclopedia with a size larger than the Library at Alexandria yet with the portability of a paperback. (The Digital Public Library of America promises to make every book ever published easily and freely accessible to anyone.) What will it mean for us to have the largest collection of knowledge in our devices, or accessed via our Google Glass? What does formal education look like when the world's knowledge is so easily accessible? What will we have to remember when so much can be easily queried?

Scenario 2: The internet develops into an autonomous 'brain,' and our brains will tightly 'interface' in a co-operative manner with this autonomous cognitive system. Futurists such as Ray Kurzweil envision a blending of the electronic brain and the biological brain (at a moment he calls the Singularity) where implants or other such devices renders the boundary between biology and technology meaningless. What will the nature of this cognitive co-operation be? What cognitive acts will we off-load onto this 'other brain?' How much of our thinking will be carried out by this exterior brain? Will this digitally enhanced mind lead to a culture that is as different as literacy was from orality?

Scenario 3: Both of the above scenarios assume a continued extension of computing power and an ever-growing capacity to off-load cognition onto digital systems. This scenario considers the potential limits and countervailing trends. Physicists have observed that we will eventually reach the physical limits of computing power (Moore's Law won't operate indefinitely since we will reach the physical limits imposed by the material objects needed to conduct computations). Futurists have warned of an impending energy shortage in electricity, such that even parts of the Western world might experience blackouts such that electricity is available intermittently. How does electronic

DOI: 10.1057/9781137460950.0002

cognitive off-loading work when electricity is so unpredictable? Others have posited sunspot and solar flare activities that could disrupt/wipeout the electronic grid, and with it the digital memory of our external symbolic storage system. How will cognition be disrupted when the electronic infrastructure of our cognition is so unstable? What happens when our digital external memory cannot function as a reliable extension of our cognition? Will we return to 'older' forms of cognitive extension? Because of its relative reliability, will we return to print as our preferred medium? Even if the system is not disrupted either by calamity or physical limits, what does 'memory' look like when digital information appears fleeting and evanescent, with a dramatically reduced shelf-life? Does the brain have a physical carrying capacity beyond which it can no longer engage in meaningful cognitive off-loading? Are their limits to our ambitions such that we may jettison our quest to develop an autonomous brain?

To understand what the future holds for the brain–Internet interface, beyond the present moment, we must place it in a much wider and deeper historical and temporal context. To do so will, I believe, make us less alarmed by the impact the Internet is having on our minds and on our culture, and more balanced in our assessments.

The goal of this interpretive essay is to grasp the meaning of the emergence, development, and trajectory of the brain–Internet interface. One of the chief attributes of the historian's skill set is that we are sense-makers: we do not simply recall events but we attempt to adjudge the broader meaning and implications of those events. Similarly, some of my favorite engagements as a futurist are when organizations ask me to help them make sense of all of the information they have swirling around them. They already know about the trends: they seek someone who can help them make sense of it all. Indeed, I considered titling this essay 'On the Meaning of the Internet,' which, perhaps can serve as a useful subtitle for what will follow. My principle activity as a humanities scholar – whether I am acting as an historian or designer or futurist – is to interpret texts.[6] A text, in my estimation, is any object created by the human brain, the most wondrous and complex object in the universe. The humanities interpret the meaning of that storehouse of human-created words, images, structures, movements, and sounds, all the texts that humans have secreted from the brain. This essay treats the Internet as a 'text,' as a symbolic object produced by the human mind, a text that can be 'read' and interpreted as any other. This essay represents my reading of the Internet, and attempts to discern the hermeneutics of the Internet.

DOI: 10.1057/9781137460950.0002

Notes

1 John Brockman, *Is the Internet Changing the Way You Think? The Net's Impact on Our Minds and Future* (New York: Harper Perennial, 2011).

2 Nicholas Carr, 'Is Google Making Us Stupid?' *The Atlantic*, July/August 2008, http://www.theatlantic.com/magazine/archive/2008/07/is-google-making-us-stupid/6868/; Nicholas Carr, *The Shallows: What the Internet is Doing to our Brains* (New York: W.W. Norton, 2010).

3 See especially the chapter 'The New Neurohistory,' in Daniel Lord Smail, *On Deep History and the Brain* (Berkeley: University of California Press, 2008), 112–156.

4 Andrew Odlyzko, 'The myth of Internet time,' *MIT Technology Review*, April 1, 2001, http://www.technologyreview.com/review/400952/the-myth-of-internet-time/

5 Fernand Braudel, 'History and the Social Sciences: The Longue Duree,' in Fernand Braudel, trans. by Sarah Matthews, *On History* (Chicago: The University of Chicago Press, 1980), 27.

6 Geoffrey Galt Harpham succinctly states that the rationale or method of the humanities is 'The scholarly study of documents and artifacts produced by human beings in the past enables us to see the world from different points of view so that we may better understand ourselves.' I would refine this even further to say that 'Humanists read texts,' where 'read' means interpret and 'texts' refer to symbolic object created by human beings. My goal here in this essay is to treat the Internet as a text, an object created by humans, and to read/interpret that text. Geoffrey Galt Harpham, *The Humanities and the Dream of America* (Chicago: University of Chicago Press, 2011), 23.

DOI: 10.1057/9781137460950.0002

Acknowledgments

For their careful reading and thoughtful comments on early versions of this essay, I thank Alexa Reck, Diane Dagefoerde, Christian Long, Stuart Hobbs, Colin Allen, Josh Sternfeld, Steven Millett, Carole Fink, Stephen Fiore, and the anonymous reviewer for Palgrave. I especially thank Jeffrey Barlow and his team at the Berglund Center for Internet Studies for the invitation to reflect on the meaning of 'Internet time' in the article 'The Internet and the Just-In-Time Mind,' which appeared in the collection *Internet 2.0: After the Bubble Burst*. Portions of the present essay are adapted from that earlier work.

DOI: 10.1057/9781137460950.0003

1
Extend

Abstract: *This chapter introduces the theory of the 'extended mind,' which holds that human cognition consists of both mental activities occurring in the biological brain in partnership with cognitive technologies outside of the brain. Unlike other species, humans have developed 'symbolic technologies' that enable us to engage in cognitive tasks that our biological brain alone would not be able to perform. The Internet represents the next stage in a long historical-evolutionary process whereby humans have expanded cognition via a symbiosis of the brain and a larger system of external, technologically enhanced memory storage.*

Keywords: brain; cognition; culture; internet; mind; symbolic

Staley, David J. *Brain, Mind and Internet: A Deep History and Future.* Basingstoke: Palgrave Macmillan, 2014. DOI: 10.1057/9781137460950.0004.

The Internet represents the next stage in a long historical-evolutionary process of expanding the capacity of the mind via a symbiosis of the brain and a larger system of external, technologically enhanced memory storage.

If one accepts this formulation, it is in no small measure due to one's acceptance of the seminal work of Andy Clark and David Chalmers. Their 1998 essay 'The Extended Mind' argued that human cognition has always consisted of both mental activities occurring in the biological brain accelerated with cognitive prostheses outside of the brain. Rejecting the Cartesian notion that cognition occurs only 'inside the skull,' Clark and Chalmers advanced what they termed 'active externalism,' meaning that the external environment (external to the brain) plays 'an active role...in driving cognitive processes.'[1] The external environment is a catch-all term for objects outside the human body that work in concert with the brain to perform cognitive tasks; these external objects are not merely aids to the brain but work in tandem with the brain to facilitate cognition. Thus, a pen and paper allow one to make complicated calculations, an example of 'the general tendency of human reasoners to lean heavily on environmental supports.' The use of slide rules, books, and diagrams are all instances where 'the individual brain performs some operations, while others are delegated to manipulations of external media.'[2]

Those manipulations of external media, not only the activities of the biological brain, constitute cognition. 'The human organism,' in this view, 'is linked with an external entity in a two-way interaction, created a *coupled system* that can be seen as a cognitive system in its own right...If we remove the external component the system's behavioral competence will drop, just as it would if we removed part of its brain.'[3] Removing pen and paper from the equation does not imply that the brain alone merely carries out the same operations: the implication here is that the level of cognition that is possible can only occur when it is coupled and amplified with the aid of the external object. I would not be able to make long extended calculations without a slide rule, my biological brain alone limits the capacity of my cognition. The addition of pen and paper or a slide rule extends my cognitive capacity; thus, the slide rule, in Clark and Chalmer's judgment, must be considered an integral part of the cognitive architecture of the mind. This is not to suggest, as critics attempted to contend, that the slide rule by itself is capable of autonomous cognition (although we will return to this point later: the stage at which our external objects start to behave in 'autonomous' ways). It is

DOI: 10.1057/9781137460950.0004

to suggest that, coupled with the brain, the tandem engages in a level of cognition not possible by either component alone.[4]

Clark and Chalmers were challenging the Cartesian view of cognition, which holds that cognition resides exclusively within the skull, and that cognition equates strictly to the activities of the brain. The assumption that the mind and all of its cognitive activities reside exclusively within the biological body, and in a highly localized portion of that body, has been influential for centuries among both philosophers and cognitive scientists. In contrast, the philosopher Mark Rowlands identifies what he terms a 'new science of the mind,' a new way of thinking about cognition 'inspired by, and organized around, not the brain but some combination of the ideas that mental processes are (1) embodied, (2) embedded, (3) enacted, and (4) extended.'[5] Rowlands' non-Cartesian cognitive science is not yet a fully developed scientific approach; he is only identifying the philosophical and conceptual outlines of this new approach to cognitive science, one based not on the idea of a mind that resides exclusively in the brain, but which has for a very long time extended outward toward an information environment.[6]

Rowlands observes that human beings often 'offload' portions of our cognitive activities to technologies residing in this external environment. His example is a GPS system or a Mapquest map, which Rowlands accesses rather than retaining spatial directions solely within his biological memory. He terms technologies like GPS systems 'external forms of information storage,' observing that they 'reduce the burden on my biological memory.'[7] But to reiterate: this off-loading of cognitive activity did not begin with Mapquest or even with the recent electronic communications revolution. The human mind has always been so extended and embodied, at least since the development of writing (although we can extend this cognitive off-loading even further in our history). It is a conceit of Western thought since Descartes that the mind is sheltered and isolated within the cathedral of the brain. The Internet, in this reading, is simply the next feature of our external environment onto which we are off-loading cognitive activity. The Internet and brain are forming yet another coupled system of cognition.[8]

Ours has long been a 'hybrid mind.' Humanity, unlike any other species, has developed symbols that take material form outside of the body, symbolic objects that allow us to store in permanent forms thoughts and ideas that we cannot store in our biological brain alone. These external forms of memory afford humans the opportunity to

DOI: 10.1057/9781137460950.0004

reflect upon, exchange, and elaborate on the thoughts and ideas so encapsulated in material form.[9]

Our 'symbolic technologies' enable us to engage in cognitive tasks that our biological brain alone would not be able to perform, either because its storage capacity is too limited or because we would not be able to conceive of a thought without the partnership of some cognitive prosthesis. The printed book, paintings, and maps are such symbolic technologies that extend our cognitive ability.[10] These symbolic technologies allow humans to escape the bonds and limitations of the nervous system, in Merlin Donald's interesting phrasing. And these technologies are not in opposition to human needs and goals, but are interwoven in the very fabric of our minds. I view the Internet as the next great symbolic technology so interwoven with the mind. I distance myself from those who would argue that the Internet is somehow robbing us of our humanity: the Internet, like all symbolic technologies, *is* our humanity.

The embodied, extended mind has long been a feature of our species. The archeologist and prehistorian Colin Renfrew argues that our encounter with the physical objects that we have fashioned from the material world has been so central to our cognition that such cultural development is what has driven our evolutionary development. Genetic change being a relatively slower process means that it alone is insufficient an explanation for the explosion of creativity and innovation that has marked our species, especially since the Neolithic Age. Indeed, Renfrew and other prehistorians have described the process of the brain coupling with objects of material culture for purposes of cognition as 'co-evolution.'[11]

Renfrew suggests that the discovery (or is it invention?) of the concept of weight provides an example of this co-evolutionary interaction between the brain and the material environment surrounding the brain. Archeologists have found among the artifacts of prehistoric sites a number of shaped objects that clearly ascend in weight according to a patterned order. Humans formed these objects, it appears, in order to discern different qualities of 'heaviness.'[12] Before the mind could conjure the concept of weight, humans first had to have a physical experience of weight, and that experiences came from the objects fashioned by human hands. That is, you would need to have a body in space that had encountered the experience of lifting a heavy object: the brain could not invent the concept without the bodily experience.

If you have such a symbolic relationship, the stone weight has to relate *to some property that exists out there in the real world.* In a sense these stone cubes serving as weights are symbolic of themselves: weight as a symbol of weight. It may be appropriate here to use the term constitutive symbol, where the symbolic or cognitive elements and the material element coexist. The one does not make sense without the other.[13] (emphasis mine)

The weights fashioned by the human hand allow the mind to engage in cognition, an interesting prehistoric example of the coupled cognitive process involved between the brain and material objects outside the brain. The archeologist Lambros Malafouris would argue that there is no Platonic concept of 'weight' waiting for the mind to perceive it. There cannot be a concept of 'weight' to be understood by the brain without the existence (manufacture) of the material objects that connote weight. There is a symbiosis here between the material object and the brain acting upon that material object: remove the material object from the equation and there is a different process of cognition. Unlike other species, humans rely on the things that we have fashioned to engage in cognition. If cognition is extended, it is extended outward into a material environment of things and objects created by the human brain. Malafouris maintains that our brains cannot function the way they do without a larger 'cognitive ecology' made up of material things, an external space that provides the context for brains to engage in cognition.[14] Indeed, he contends that the brain – 'seen as internal assemblies of neurons' – and culture – 'seen as external assemblies of material structures and scaffoldings' – by themselves are 'lifeless.' Both come to life only when in interaction with each other.[15] Early in prehistory – a very long time before the emergence of the Internet – humans were developing an intimate relationship with things of their own creation as a way to engage in cognition.

The human mind cannot conceive of the concept of weight without an external 'thing to think with.'[16] Those weights are not 'natural' objects, but are rather fashioned by humans. This is a theme that we will continue to explore: that the human mind conceives of things and objects with which to engage in cognition, without which the brain would not be able to engage in the cognitive act. That feedback loop between brain, object, and cognition is a complex, emergent cognitive process that has driven our (cultural) evolution.[17] The brain + external objects allow humans to engage in cognitive processes not possible via the brain alone. It is in this context that I wish to understand the historical-evolutionary place of the

DOI: 10.1057/9781137460950.0004

Internet: as the (next) thing we have added to our external environment that thereby alters and expands our cognitive capacity.

Human evolution is no longer driven by Darwinian biological/genetic change alone. Indeed, our evolution was accelerated once we developed the capacity to create external symbolic storage systems. These external symbols, which Merlin Donald terms exograms, work in tandem with our biological memories, or engrams.[18] Our evolution has been determined by the growth in the quantity of exograms at our disposal, and the emergent properties of the interaction between exograms and engrams.[19]

Donald argues that 'The modern human mind evolved from the primate mind through a series of adaptations, each of which led to the emergence of a new representational system.'[20] Representational here means that humans, unlike other species, have devised ways to represent reality in tangible form exterior to what could be stored within their biological brains. Those representational systems are not supplanted by subsequent systems, but are, rather, layered upon each other.[21] These representational systems, are the key to understanding the evolution of the modern mind; any expansion in our cognitive abilities (from an evolutionary point of view) has derived from our success in developing new representational systems.[22] These systems have surrounded human beings to such an extent that 'the structure of the primate mind was radically altered; or rather, it was gradually surrounded by new representational systems and absorbed into a larger cognitive apparatus.'[23] We typically identify those larger representational systems – art, writing, music, and so on – as products of human culture. Psychology, among other disciplines, has taught us to make a separation between the individual mind and the larger 'culture' in which that mind resides. Donald, however, wants us to see the two as linked together in a single 'cognitive apparatus.' Where psychology would define the mind as existing within the physical apparatus of an individual human brain, Donald extends that definition to include the external representational systems that humans have invented. One important implication of this theory is that it joins together the realm of individual human psychology with human culture in a complex interplay that defines our species' cognitive architecture. It is within this context that I would like to understand the long-term historical significance of the Internet and its relationship to our human mind.

The early hominid brain, like all primate brains, was episodic, the kind of mind characterized by our closest evolutionary kin, the great apes.[24]

Most mammals have developed 'procedural' memories as well, meaning memories of a concrete set of actions that might be generalized over time and space. If episodic memories refer to specific events, then procedural memories are generalizations of events. Neither form, however, involves a system of signs that can be used to reflect on these events, or to pass the memories of these events on to others.[25] Only humans, it seems, have developed 'semantic' memory, which are memories captured with signs or in some other representational form.[26]

Homo erectus developed the capacity to exchange information through gestures, hand signs, facial expressions, and other extra-linguistic forms of communication that Donald describes as mimetic. *Erectus*, of course, was a tool-making species, and we must remember that to manufacture such tools required a system of communication more elaborate than an episodic mind could develop. *Erectus* did not possess language, but did develop mimetic skills, which means 'the ability to produce conscious, self-initiated, representational acts that are intentional but not linguistic ... mimesis is fundamentally different from imitation and mimicry in that it involves the invention of intentional representations. When there is an audience to interpret the action, mimesis also serves the purpose of social communication.'[27] Acts of mimesis are expressions of thought without language; indeed, before humans could develop language, before we had anything to say, 'there had to be some sort of semantic foundation for speech adaptation to have proven useful, and mimetic culture would have provided it.'[28] At this stage in human cognitive evolution, represented by mimetic culture, we are still referring to representational systems of thought that remain grounded in the human body, without material foundation. Facial expressions and ritualized dance did not yet exist as physical, tangible objects external to the human body.[29] Nevertheless, 'the brain structures supporting mimetic action ... constituted the archaic human brain, the brain that would be *further modified* to incorporate linguistic skill into its armamentarium of systems and modules'[30] (emphasis mine). The emphasis here is to draw attention to the long history of brain modifications that have defined the evolution of the human mind.

Modern *homo sapiens* added linguistic ability and greater 'semantic skill' to this growing cognitive architecture. With the development of spoken language, 'the human mind had come full circle, starting as the concrete, environmentally bound representational apparatus of episodic culture and eventually becoming a device capable of imposing an interpretation of the world from above'[31] through language. Humans built upon the episodic and mimetic mind by weaving stories and myths that would bring order to the world through language.

Importantly, and especially for our understanding of the possible effects of the Internet both on the human mind and human culture, mimetic skill and mimetic culture did not 'disappear' when humans developed more advanced cognitive systems. Aspects of mimetic thought – vocal tone, mime, facial expression and gesture, eye movement, sport, and other ritualized movements – remain a vital part of our cognitive architecture. Donald describes mimetic culture as 'vestigial,' meaning that mimetic skill was not lost when newer, more advanced forms of thought and communication were developed. Rather, such skills were enveloped by the new culture, an additive process. Indeed, this schema rests on the idea that these earlier forms of representation remain embedded within the larger human cognitive architecture. 'Episodic culture,' for example, 'would have been surrounded by, *and largely preserved*, within the larger context of mimetic culture... the transition to mimetic culture involved adding to the cognitive architecture already in place'[32] (emphasis added). For those who wonder about the fate of the Gutenberg Galaxy and of typographic thought and culture in the Age of the Internet, it may be useful to think of written culture as potentially vestigial to that culture, meaning that it will be embedded within – but not eliminated by – that culture.

The episodic and mimetic minds remained biologically-based. That is, whatever systems of representation humans developed, these older systems remained tied exclusively to the human body; the representational signs and symbols were contained in the physical, biological apparatus of the human brain, and were expressed through the body. This suggests that any information and communication was time- and space-dependant, because eye movements or ritualized dance or hand signs were not captured in material form, any thought or communication expressed was evanescent. Thus, an important cultural shift occurred around 40,000 BC, when humans began to create 'tokens of memory' that gave physical form to thoughts, ideas, and information. Earlier humans were decorating their bodies with paint, and these decorations more than likely carried cultural meanings. But when humans began carving Venus figurines and painting animal figures on cave walls, we took an important evolutionary step in that our mimetic gestures were now preserved outside the human body, outside the biologically based brain.[33]

DOI: 10.1057/9781137460950.0004

The preservation of thought and memory in external material form advanced in turn. After painting on walls, humans developed pictographic and hieroglyphic writing systems and, as speech and visual symbols were united, the first alphabetic writing systems. We developed token-counting and other systems of account, musical notation, mathematics, all of which are based on creating visual representations of signs and symbols. Once preserved outside the body, these visual and material representations could be looked at, shared with others, reflected and commented upon, and, importantly, 'remembered' without the reliance on biological memory. It is often said of the first counting systems, for example, that as long as what needed to be counted was small in number, what needed to be remembered could be contained in the biologically based human memory. However, as societies grew larger and more complex – that is, as more objects needed to be counted and remembered – a system of visual, material numerical notation was necessary to supplement human memory. This supplement to memory should not be distinguished from our biologically based memory. Indeed, it should be viewed as an extension of that memory, our biology extended outward through our technologies.

Donald refers to this ever-expanding corpus of material representations the 'external symbolic storage system.' This concept pertains to 'all memory items stored in some relatively permanent external [to the human brain] format, whether or not they are immediately available to the user.'[34] What we identify as the products of human culture and civilization might be better understood as extensions of the human mind, inseparable from that biologically based entity. This external symbolic storage system has surrounded human beings since the beginnings of civilization, with additions over the millennia. All the objects we associate with culture – art, music, architecture, as well as books and, it must be said, the Internet – are contained in this representational storage system, a 'cloud' of symbols that surrounds each individual human mind.

Interestingly, Donald evokes the metaphor of the computer to conceptualize this human mind defined by both its internal biologically based memory and its external memory. The computer clearly has its own hardware and software stored on the client, but computers also rely on external memory devices (like a USB drive or portable hard drive) that extend the functionality of the computer beyond that stored on the client. (Indeed, computer specialists speak of 'external memory' in this way.) Donald was writing before the rise of the World Wide Web, but understood the implications of networked computing, especially its ability to extend the

DOI: 10.1057/9781137460950.0004

functional power of the individual computer. 'If a computer is embedded in a network of computers, that is, if it interacts with a "society" of other computers, it does not necessarily retain the same "cognitive capacity." That is, the powers of the network must also be taken into account when defining and explaining what a computer can do.'[35] Similarly, 'individuals in possession of reading, writing, and other visuographic skills thus become somewhat like computers with networking capabilities; they are equipped to interface, to plug into whatever network becomes available. And once plugged in, their skills are determined by both the network and their own biological inheritance.'[36] Later in this essay, we will consider the future of our interface with the Internet, but at this stage I wish only to point out that 'interface' with our symbolic technologies is a long-standing feature of our species: that to engage in cognition means to interface with our storehouse of external symbols. At one time, 'plugging into' our cognitive network meant interacting with a book or viewing a painting. Today, 'plugging into' our collective external symbolic storage system also means plugging into the Internet.[37]

In order to grasp its meaning – both its place in the larger historical-evolutionary narrative and its possible future directions – I would like to place the development of the Internet within this much longer deep history, and contend that the Internet must be considered an important part of the human mind that is 'out there,' the next step in the longer historical process of interacting with symbols, information, culture, and memory stored in material form outside of our bodies. What I describe as 'just-in-time' knowledge should be understood within this long-term historical process. More and more information and knowledge is migrating to the Internet and even more is being created in situ. I am reminded of the doctor I saw once in an emergency room, consulting a pocket version of his Physicians' Desk Reference in order to check on potential risks of prescribing two drugs. Doctors are not expected to keep all of this information in their heads, and thus the need for a pocket 'memory device.' Of course, those pocket editions of the Physicians' Desk Reference are now found on smart devices, meaning that it is even easier for doctors to access that information 'just-in-time.'

The 'cloud' is the ideal metaphor for this historic development. In computing terms, the Cloud refers to data that is stored on an external server, as opposed to on a local client device. It is exactly what I have in mind in describing just-in-time knowledge: rather than relying solely on internal, biological cognitive processes, we have devised a new way of accessing and manipulating symbols from an external storage medium. The Internet provides us with even greater storage capacity than any other symbolic technology so created. It is a 'thing' that permits us to engage in accelerated forms of thought and cognition. The Internet is the most portable symbolic technology yet developed, allowing us to, in effect, carry around entire libraries in our pockets.

For as long as we have had symbolic technologies and 'things to think with,' we have been surrounded by a symbolic 'cloud.' I take the step of delving into the deep historical-evolutionary past as a way to contextualize our current Internet moment, to suggest that extending our cognitive capacities through external symbolic technologies is an ancient practice and a natural human impulse.

Notes

1   Andy Clark and David J. Chalmers, 'The Extended Mind,' in Richard Menary, ed. *The Extended Mind* (Cambridge, MA: MIT Press, 2010), 27.
2   Ibid., 28.
3   Ibid., 29.
4   For a thoughtful critique of the extended mind hypothesis, see Robert D. Rupert, *Cognitive Systems and the Extended Mind* (New York: Oxford University Press, 2009).
5   Mark Rowlands, *The New Science of the Mind: From Extended Mind to Embodied Phenomenology* (Cambridge, MA: MIT Press, 2010), 3.
6   Rowlands, *The New Science of the Mind*, 3. 'The idea that mental processes are embodied is, very roughly, the idea that they are partly constituted by, partly made up of, wider (i.e. extraneural) bodily structures and processes. The idea that mental processes are embedded is, again roughly, the idea that mental processes have been designed to function only in tandem with a certain environment that lies outside the brain of the subject.... The idea that mental processes are extended is the idea that they are not located exclusively inside an organism's head but extend out, in various ways, into the organism's environment.'
7   Rowlands, *The New Science of the Mind*, 14.
8   Harry Halpin, Andy Clark and Michael Wheeler, 'Towards a Philosophy of the Web: Representation, Enaction, Collective Intelligence,' Proceedings of the Web Science Conference: Extending the Frontiers of Society On-Line, April 26–27, 2010. http://citeseerx.ist.psu.edu/viewdoc/download;jsessionid=DCBD0BC4BD08A6E2E4DF3A2EDCFBC023?doi=10.1.1.415.282&rep=rep1&type=pdf

9   Merlin Donald, *A Mind So Rare: The Evolution of Human Consciousness* (New York: W.W. Norton, 2001), 305.

10   Donald, *A Mind So Rare*, 305.

11   Colin Renfrew, *Prehistory: The Making of the Human Mind* (New York: The Modern Library, 2007), 80.

12   Renfrew, *Prehistory*, 99–100. 'When a series of well-shaped objects made of dense material and of ascending size, discovered among the artifacts from some prehistoric culture, are today weighed and found to be multiples of what we would call a unit of weight, it is often reasonable to infer that the culture in question had formulated its own system for units of mass.

But if we go on to ask what these new artifacts were symbols of, it turns out that they were used to symbolize and quantify an inherent property not previously identified or quantified, which then became isolated for study and measured for the first time. We are today all familiar with the notion of weight, both as a measure and as utilizing the simple idea that something may have weight and be heavy. But it is worth considering how the notion of measurable weight could come about in the first place.'

13   Renfrew, *Prehistory*, 100.

14   Lambros Malafouris, *How Things Shape the Mind: A Theory of Material Engagement* (Cambridge, MA: MIT Press, 2013), 245.

15   Malafouris, *How Things Shape the Mind*, 84.

16   This term comes from Esther Pasztory by way of Claude Levi-Strauss, who stated, according to Pasztory, that 'things are good to think with, rather than merely good to look at.' She writes: 'It would seem that the thinking process needs projections on and manipulation of things to work itself through to consciousness or to demonstrate itself to itself. In that sense it could be said that [picturing arranging or closet arranging is] a form of magic ritual. I reorganized the world by manipulating symbols. My means were aesthetic (what looked right), but my urgent concern was cognitive (what was right and matched certain sets of data). Things are needed to think with, in order to manage problems of cognitive dissonance.' Esther Pasztory, *Thinking With Things: Toward a New Vision of Art* (Austin: University of Texas Press, 2005), 21.

17   Renfrew, *Prehistory*, 101–102.

18   Donald, *A Mind So Rare*, 309.

19   Malafouris, *How Things Shape the Mind*, 84. 'The cognitive life of things, like the cognitive life of brains, can be found where engrams and exograms begin spiking, interacting, and complementing one another in such a way that memory emerges.'

20   Merlin Donald, *Origins of the Modern Mind: Three Stages in the Evolution of Culture and Cognition* (Harvard University Press, 1991), 2.

21   Donald, *Origins of the Modern Mind*, 2–3. 'Each successive new representational system has remained intact within our current mental architecture, so that the modern mind is a mosaic structure of cognitive vestiges from earlier stages of human emergence.'

22   Donald, *Origins of the Modern Mind*, 3. 'Humans did not simply evolve a larger brain, an expanded memory, a lexicon, or a special speech apparatus; we evolved new systems for representing reality.'

23   Donald, *Origins of the Modern Mind*, 4.

24   Ibid., 149. 'Their lives are lived entirely in the present, as a series of concrete episodes, and the highest element of their system of memory representation seems to be at the level of event representation.'

25   Donald, *Origins of the Modern Mind*, 151.

26   Ibid., 160.

27   Ibid., 168.

28   Donald, *Origins of the Modern Mind*, 199. Steven Pinker has argued that, contrary to a long tradition in linguistics, thought does not equate to language. Indeed, he has posited a quality he calls 'mentalese,' thoughts that exist before they are represented in language or any other symbolic system. He observes: 'We have all had the experience of uttering or writing a sentence, then stopping and realizing that it wasn't exactly what we meant to say. To have that feeling, there has to be a "what we meant to say" that is different from what we said.' See Steven Pinker, *The Language Instinct: How the Mind Creates Language* (New York: Harper Perennial, 1995), 57.

29   On the relationship between gesture and thought, see David F. Armstrong, William C. Stokoe and Sherman E. Wilcox, *Gesture and the Nature of Language* (Cambridge: Cambridge University Press, 1995); and David McNeill, *Hand and Mind: What Gestures Reveal About Thought* (Chicago: University of Chicago Press, 1992).

30   Donald, *Origins of the Modern Mind*, 200.

31   Ibid., 268.

32   Ibid., 197.

33   For an interesting history of the idea of information, see Michael E. Hobart and Zachary S. Schiffman, *Information Ages: Literacy, Numeracy, and the Computer Revolution* (Baltimore: Johns Hopkins University Press, 1998). The authors argue that 'the invention of writing gave birth to information itself, engendering the first information revolution. Writing created new entities, mental objects that exist apart from the flow of speech, along with the earliest, systematic attempts to organize this abstract mental world.' (2) It was the action of preserving human thought in material form that constituted 'information.' The authors write 'Both writing and speech constitute communication, but of the two only writing extracts the sounds of speech from their oral flow by giving them visual representation ... Because

information separates mental objects from the flux of experience, it follows that different information technologies can single out different aspects of experience in different ways, generating different kinds of information' (4–5). This analysis strikes me as very similar to what Donald was suggesting about the movement toward semantic representation and the creation of the external symbolic storage system. Hobart and Schiffman see writing as the prime mover event in the creation of information, but, as I note above, I see this invention with the first stone figurines and cave paintings. To follow their definition, these were mental objects separated from the flow of experience. However we term it, there was an historical development from evanescent forms of communication that were not preserved in material form versus those that were so preserved.

34   Donald, *Origins of the Modern Mind*, 306.

35   Ibid., 310.

36   Ibid., 311.

37   Donald uses this metaphor of the computer to describe the human mind as it has developed since the Neolithic. Interestingly, his metaphor is not necessarily that shared by some AI researchers, who have long sought to replicate the mind in the computer, or at least to view the mind as a kind of computer. Rather, Donald's analogy is the inseparability of internal and external memory storage as a way to understand the computational power of the entire computer. In the same way that we cannot separate the totality of the computer's processing power into hardware and software stored in the machine, on the one hand, and its connection to other machines via a network on the other, neither can we separate the biological portion of the human mind from the larger 'external storage system' humans have developed to extend that mind. The products of human culture must be viewed as 'hardware' every bit as much as the biologically based brain is hardware.

DOI: 10.1057/9781137460950.0004

2
Reconfigure

Abstract: *Is the Internet making us stupid? There is a growing body of research that suggests that the Internet is rewiring the synaptic patterns of our brains, which for some is a cause for alarm. All symbolic technologies – not just the Internet – have rewired our brains, and thus the Internet is unremarkable in this way. In particular, the Internet appears to encourage associative and analogical – rather than linear and logical – thinking. The brain is indeed linear and logical, but the brain has also proven to be analogical and associative, capable of making connections between disparate objects and data points, and has long been doing so. The Internet has not dulled our minds but has instead unleashed this pre-existing, if undervalued, portion of our cognitive architecture.*

Keywords: brain; culture; Internet; mind; reading, symbolic; writing

Staley, David J. *Brain, Mind and Internet: A Deep History and Future.* Basingstoke: Palgrave Macmillan, 2014. DOI: 10.1057/9781137460950.0005.

If the mind comprises both the physical brain inseparable from the larger external symbolic storage system, then it would seem that changes to that external system have important implications for the mind. To that end, I am interested in the line of thought begun by Nicholas Carr, who asked in a widely read article (and later extended in a thoughtful book) whether 'Google is making us stupid.'[1] His concern is that how we use the Internet is dulling our capacity for deep reading, and, thus, deep thought and reflection. Users of the Internet do not so much read as flit from hyperlink to hyperlink, voraciously consuming content, but not digesting it with the patience and thoughtfulness that we associate with reading books or long-form articles. More worrying for Carr is that the Internet is, perhaps without our knowledge and consent, rewiring our brains.

Carr's is an important argument, and is not the standard way we think about the impact of new technologies or new media. Many commentators have documented how new tools change our culture: the development of the mechanical clock, for example, altered our natural biological rhythms and enforced on humanity a 'clock culture' of mechanized time. According to economic historians, the Industrial Revolution required of workers less artisanal skill and more 'machine tending,' a de-skilling of their labor. What is new in Carr's estimation is that we now understand more about how these tools directly affect the synaptic patterns in our brains. Carr cites important new work that measures the ways in which prolonged Internet use creates new synaptic connections, literally rewiring our brains. If this new research demonstrates anything, it is that the brain is a far more plastic organ than we previously imagined. And because we can now measure the extent of the changes wrought by this new technology using brain scanning technologies like fMRI (functional Magnetic Resonance Imaging), it seems all the more dastardly and insidious. Technophobes and Luddites have long decried the deleterious effects of technology on human culture and society; it would seem that these critics have a new – and, ironically, technologically determined – way to legitimate their concerns. As Carr notes, 'for all that's been written about the Net, there's been little consideration of how, exactly, it's reprogramming us.'[2]

Carr cites in evident agreement the thoughts of Richard Foreman, who elegizes the end of the individual personality in a culture wired to the Internet. 'I come from a tradition of Western culture in which the ideal (my ideal) was the complex, dense and "cathedral-like" structure of the highly educated and articulate personality – a man or woman who carried inside themselves a personally constructed and unique version of the entire heritage of the West,' begins Foreman.

> But today, I see within us all (myself included) the replacement of complex inner density with a new kind of self – evolving under the pressure of information overload and the technology of the 'instantly available.' A new self that needs to contain less and less of an inner repertory of dense cultural inheritance – as we all become 'pancake people' – spread wide and thin as we connect with that vast network of information accessed by the mere touch of a button.[3]

I cannot read Foreman's thoughtful quote without thinking of the long history of the external symbolic storage system. We have always – at least as long as we have been *homo sapiens* – connected to a vast network of symbols residing outside the brain. It is just that the scope and nature of that network and the manner in which we connect to it has changed over time. I am reminded of the image of 'St Jerome in his Study,' one of the iconic visual metaphors of Western culture. In various visual representations, the great scholar sits alone among his books, globes and calipers. Far from being alone within the cathedral of his mind, I view those portraits of St Jerome as a man whose mind consists of both his brain and the technical apparatus of his study. To what degree is the self that is 'St Jerome' indistinguishable from his study? How has the particular configuration of his books defined his mind? With each book added to the shelf of his study, with each book that he reads or writes, how has the definition of his 'inner self' been altered? Can St Jerome wall off his mind from the cognitive prosthesis of his study? Is the scholar defined in part by his books? Can there be a St Jerome without his study? Given the importance of the external symbolic storage system in defining our cognitive capacities, it seems that we have never been able to so fully separate ourselves, to isolate our mind from our information storage networks. Is this notion of an isolated self a Western conceit?

It seems that the concern expressed by Carr and Foreman and others is that we lack autonomy from our technologies. Considered in terms over the historical long-view, the human mind has never been as isolated as Foreman supposes. He concludes his reflections with a question: 'Can computers achieve everything the human mind can achieve?'[4] But framing the question in this way again assumes that our information technologies are as autonomous as the self supposedly is. This is a typical

DOI: 10.1057/9781137460950.0005

rhetorical move, I think: we position technology – and especially new media – as in opposition to humanity, the Other that enslaves us (if we are a Luddite) or liberates us (if we are a technophile).

This view of technologies as exogenous to humanity holds whether one is a technological determinist or a technological instrumentalist. Carr sides unequivocally with technological determinists, who maintain that humans are shaped by their tools; an instrumentalist holds that tools only do what we tell them to do.[5] Whether one is a determinist or an instrumentalist, however, the assumption in both cases is that technologies exist apart from humanity. This is not even a false dichotomy: the assumption is inherently flawed, especially with regard to technologies of the external symbolic storage system. Claiming that technology is exogenous to humanity is like saying that the shell is exogenous to the snail, and that the shell has its own intentions that are in some fashion 'in opposition' to the snail. Stated another way: is a snail still a snail without its shell? Is a spider still a spider without its web? Is a human being still human without cognitive technologies? More specifically, is the mind still the mind without its external symbolic storage system?

Carr reveals much when he makes the following observations:

> Language itself is not a technology. It's native to our species. Our brains and bodies have evolved to speak and hear the words. A child learns to talk without instruction as a fledgling bird learns to fly. Because reading and writing have become so central to our identity and culture, it's easy to assume that they, too, are innate talents. But they're not. Reading and writing are *unnatural acts*, made possible by the purposeful development of the alphabet and many other technologies. Our minds have been taught how to translate the symbolic characters we see into the language we understand. Reading and writing require schooling and practice, the deliberate shaping of the brain.[6] (emphasis mine)

Curiously, Carr expresses little concern throughout his book for the 'unnaturalness' of reading and writing and the book culture he elegizes. (The Internet, and the undesirable effects on our brains, would seem to be more 'unnatural' than reading and writing.) Technological enhancement, in this view, is not natural, not native to the human species. One could contend that humans have not been 'natural' for millennia, but I would rather say that enough evolutionary time has passed that our external cognitive extensions are as 'natural' to us as our capacity for language.

A 'third way' is clearly preferable, a view of technology neither as an external imposition on the natural brain nor merely a neutral tool.

DOI: 10.1057/9781137460950.0005

Technologies are natural extensions of the brain; they are something humans develop because we are human, and our humanity is diminished were we left without technologies. To cease using alphabets or mathematical symbols or paintings would be the same as asking one of us to remove a limb or to blind ourselves. Tools are not exogenous to humanity, but co-evolutionary with humanity.

Katherine Hayles refers to this process as *technogenesis*, 'the idea that humans and technics have coevolved together.'[7] Hayles argues that we are witness to such a moment of co-evolutionary change when we observe synaptic alterations in the brain brought about by increasing use of the Internet. The Baldwin effect suggests that 'when a genetic mutation occurs, its spread through a population is accelerated when the species reengineers its environment in ways that make the mutation more adaptive.' That is, changing the environment – changing the cultural environment, the external symbolic storage system – can lead to evolutionary/genetic change. 'Epigenetic changes in human biology,' notes Hayles, 'can be accelerated by changes in the environment that make them even more adaptive, which leads to further epigenetic changes.'[8]

In contrast to the breathless and urgent appeal made by critics who fret about the changes wrought to the brain by increasing use of the Internet, Hayles affects a more neutral, matter-of-fact statement about these changes:

> As digital media, including networked and programmable desktop stations, mobile devices, and other computational media embedded in the environment, become more pervasive, they push us in the direction of faster communication, more intense and varied information streams, more integration of humans and intelligent machines, and more interactions of language with code. These environmental changes have significant neurological consequences, many of which are now becoming evident in young people and to a lesser degree in almost everyone who interacts with digital media on a regular basis.[9]

Hayles agrees that the Internet is changing our brains, but that that epigenetic change in the cultural environment has been created by humans. This is not an unnatural, autonomous force but a product of the active and ever-generative human brain. This is a recursive process: changes in the material environment come from the very brain that is being altered by those environmental changes.

I wonder if the terms of this discussion would be altered if, instead of saying 'technology,' we used the term 'culture?' In the language we have

DOI: 10.1057/9781137460950.0005

been using here, the products of the external symbolic storage system are what we usually call 'culture.' I suppose that I have been using the term 'technology' here (perhaps because I am following Carr's line of thinking) but it seems to me that, at least when discussing the technologies of mind and external storage, we are talking about culture. As a humanist, I cannot seriously contemplate the idea that 'culture' is external to and in opposition to humanity. Our culture is our humanity. The deeper history of civilization, however, suggests that symbolic technology – that secretion of the human mind – is intertwined with the mind, not in separation and isolation. The Internet can only change what we can change with them in co-evolutionary partnership.

Foreman fears a 'flattening out' of the self. This implies that the individual mind is being hollowed out as it no longer needs to serve as a repository of knowledge. Memory, the storage of knowledge in the individual mind, has always been a feature of humanity, of course. We are still, at our core, episodic and mimetic creatures, with biological memories. But we have never been so fully dependent upon that memory, that inner cathedral, as Foreman might have imagined it, even, I would contend, during periods where memory was highly prized – among oral cultures, say, or among the ancient Romans – the human mind has long depended upon external systems to expand its memory, its capacity. The Internet, in one telling, is emptying us of our knowledge, but only if we were to conceive ourselves as cut-off from – in opposition to – that external system.

I wonder if the concern that the Internet is rewiring our brains is implying that the brain is a pristine organ that has never been rewired before, has never been so 'violated' by our technologies? We should point out that our external symbolic storage system has historically had a similar effect on our brains. Indeed, 'things change the brain,' states the archeologist Lambros Malafouris, suggesting that all of the objects fashioned by the human brain since prehistory have altered that brain. Things 'effect extensive structural rewiring by fine tuning existing brain pathways, by generating new connections within brain regions, or by transforming what was a useful brain function in one context into another function that is more useful in another context.'[10] The cognitive neuroscientist Stanislas Dehaene concludes that the invention of writing repurposed the human brain for the task of reading the written symbols. Our brains are evolutionarily not far removed from our primate brains, Dehaene argues, and thus humans developed a 'reading brain' by reordering a brain that had developed for other purposes. 'If books and libraries have played a predominant role in the cultural evolution of our species,' he writes, 'it is because brain plasticity allowed us to recycle our primate visual system into a language instrument. The invention of reading led to the *mutation* of our cerebral circuits into a reading device'[11] (emphasis mine).

Dehaene advances what he calls the neuronal recycling hypothesis. 'According to this view, human brain architecture obeys strong genetic constraints, but some circuits have evolved to tolerate a fringe of variability. Part of our visual system, for instance, is not hardwired, but remains open to changes in the environment. Within an otherwise well-structured brain, visual plasticity gave the ancient scribes the opportunity to invent reading...When we learn a new skill, we recycle some of our old primate brain circuits – insofar, of course, as those circuits can tolerate the change.'[12] It would seem that 'changes in the environment' means especially the cultural/informational environment, an environment of our making that, in turn, requires our brains to be recycled and reordered. If such a process defined the origins of writing, there is every reason to think that a similar process is ordering our brain's adaptation to the Internet.

We see evidence of this reordered and repurposed brain as we observe children learning to read. In miniature, and with brain-imaging tools, we can witness the recycling/repurposing process, the 'mutation' process, at work.

> If one could zoom down to the scale of single neurons or cortical columns, one would see a major upheaval in the neuronal microcode. According to the recycling view, each reading lesson leads to a neuronal reconversion: some visual neurons, previously concerned with object or face recognition, are committed to letters; others to frequent bigrams; yet others to prefixes, suffixes, or recurring words. In parallel, the neural code for spoken language is also in flux. Somehow, as phonemic awareness emerges, the code explodes into a more refined structure where phonemes are explicit. Finally, if we could track nerve fibers during development and sort them out depending on function, we would see a regular, comblike projection appear that links each visual unit to its corresponding pronunciation.[13]

The acquisition of reading alters and rewires the brain. Children learning to read reenact an ancient process of repurposing the brain. I suspect that if we had access to fMRI devices when the first writing systems were developed, we would probably see evidence of brain rewiring at work.

DOI: 10.1057/9781137460950.0005

As part of an historical process, the Internet, like all the creations of our external symbolic storage system, reorders the brain.

For Dehaene, the brain is not infinitely plastic, only selectively plastic. That is, we developed our ability to read by altering those portions of our brain that were prepared to be refashioned for other purposes, like portions of our visual system that could be repurposed to understand written signs. These areas of plasticity are limited, and thus put constraints on the kinds of writing systems that could emerge. Dehaene notes that, despite their seeming variety and diversity, all of the writing systems humans have developed are morphologically very similar. He notes, for example, that the signs of written systems, from the alphabet to Chinese symbols, are very similar in size, in the kinds of stroke marks used to create each symbol, and so on. He contends that this is because of the structural limitations imposed by our brains; our writing systems were limited by what our brains were able to manage given its genetic makeup and its limited plasticity.[14] It strikes me that this has important implications for how we might understand the future evolution of the Internet. While we might fuss over the Internet's impact on our brains and fret about how it is uncomfortably rewiring them, we might pause to consider how our own cognitive architecture is setting limits on how the Internet is and will develop. If the Internet fosters just-in-time knowledge, then this is, in part, because our brains allow it to do so.

Perhaps the issue here is a discomfort with the uncertain direction that rewiring is leading us. Carr, especially, laments the decline of reading in depth, that our brains may appear unable to sustain thought, but rather flits around like a gnat from this bit of data to this. The judgment that the Internet is 'making us stupid' is based upon the idea that the book-reading brain is the ideal brain. Although it is so frequently evoked in such circumstances that I hesitate to do so again, I feel I must draw attention to the oft-quoted dialogue between Socrates and Phaedrus that spells out the former's objection to writing, in the form of a parable:

> But when they came to letters, This, said Theuth, will make the Egyptians wiser and give them better memories; it is a specific both for the memory and for the wit. Thamus replied: O most ingenious Theuth, the parent or inventor of an art is not always the best judge of the utility or inutility of his own inventions to the users of them. And in this instance, you who are the father of letters, from a paternal love of your own children have been led to attribute to them a quality which they cannot have; for this discovery of yours will create forgetfulness in the learners' souls, because they will not

use their memories; they will trust to the external written characters and not remember of themselves. The specific which you have discovered is an aid not to memory, but to reminiscence, and you give your disciples not truth, but only the semblance of truth; they will be hearers of many things and will have learned nothing; they will appear to be omniscient and will generally know nothing; they will be tiresome company, having the show of wisdom without the reality.[15]

Socrates feared that, in relying on an external system of symbolic storage, the young will lose the ability to remember with their biologically-based memories. If I might baldly paraphrase him, was Socrates asking 'Is writing making us stupid?'

Carr gives eloquent expression to the real fear of the loss of reading (books) as a central cognitive activity in an Internet-saturated culture. As an avid reader myself, I share this lamentation, but I should also hasten to point out that deep reading is but one way humans read. The digital humanist Matthew Kirschenbaum asks 'What is reading?' and correctly observes that we engage in many types of reading. How we read a novel, for example, is different from how we read an encyclopedia, from how we read a poem, from how we read a recipe.[16] Indeed, the 'flitting about' process of reading seems to predate the Internet, but is certainly made much easier by the Internet. Kirschenbaum evokes images of medieval scholars in their studies with devices that held open several books at once so that the reader might read between them simultaneously, or Thomas Jefferson's device that held multiple books open at once, with Jefferson's leaping between them. 'Books are random access devices par excellence, "concludes Kirschenbaum," and the strict linear sequences of reading we associate with sitting under the tree [becoming "lost" in a book] is the exception, not the rule.'[17] Even if we are losing the ability to read deeply – and I am not yet convinced that this is the case – we could argue that what is being lost is but one type of reading among many different variations.

We might also wonder if deep reading – and the brain structure that supports this activity – is becoming a vestigial part of our minds. That is, deep reading is not lost so much as it is subsumed under the new brain activities fostered by the Internet. I understood Donald's use of the word *vestigial* here not to mean 'leftover and without apparent use,' like the human appendix, but rather a trace of something older. He described mimetic skill, such as facial expressions and gestures, as vestigial in this way, as older skills that were long supplanted by vocal language and written signs, but that are nevertheless retained as important forms of communication. Language and writing were layered upon gesture, but this did not eliminate gesture from the architecture of the human mind. Similarly, I can envision a scenario where deep reading does not disappear; rather the kind of thought encouraged by the associative 'leaping about' across the Internet is layered over the deep reading apparatus. Historically, the architecture of the mind accumulates rather than eliminates.

The Internet does indeed enforce a kind of mental leaping about between bits of data. Our minds seem to flit about like a gnat when we are surfing this new external symbolic storage system. The Internet would appear to be structured to function in a nonlinear and associative manner, in contrast to a book which is organized to be linear and logical (even if it is not always read in that fashion). One of the chief concerns about the Internet and the electronic communications system generally is that it stands in contrast to The Book, which, since the development of the printing press, has served as Western culture's symbolic representation of the human mind. As J. David Bolter writes:

> Because writing is such a highly valued individual act and cultural practice, the writing space itself is a potent metaphor. In the act of writing, the writer externalizes his or her thoughts. The writer enters into a reflective and reflexive relationship with the written page, a relationship in which thoughts are bodied forth. Writing, even writing on a computer screen, is a material practice, and it becomes difficult for a culture to decide where thinking ends and the materiality of writing begins, *where the mind ends and the writing space begins.* With any technique of writing – on stone or clay, on papyrus or paper, and the computer screen – the writer may come to regard the mind itself as a writing space. The behavior of the writing space becomes a metaphor for the human mind... [18] (emphasis mine)

It is unsurprising to me that Bolter finds the boundary between the mind and the writing space difficult to discern; in the language we have been using here, we indeed cannot separate the biologically-based mind from its systems of external representation. Whenever we write, we preserve our thoughts in external symbolic form. That external writing space is not a neutral medium for our thoughts, however. Depending on whether we are writing on clay tablets or papyrus scrolls or upon the pages of a book, the material surface of our writing space shapes our thoughts. As we have noted earlier, these different reading and writing practices have real effects – not just metaphorical ones – on the physical structure of our brains.

DOI: 10.1057/9781137460950.0005

So, what sort of writing space does the Internet represent? And if our writing spaces are metaphors of the mind, what sort of mind does the Internet represent? If the Book is a writing space that fosters linear and logical thinking, then the Internet would appear to encourage associational and analogical thought. If the metaphor of the mind represented by the Internet is best described as leaping between texts in a nonlinear fashion, then we should recall that this style of reading long predates the rise of the Internet. Thomas Jefferson built a device that held open several books at once for him to swivel between; the Internet makes it easier for us to 'swivel between' a potentially infinite number of texts. The connections we make when we read in this fashion are just as likely to be associational and analogical rather than linear and logical.

Associational thinking has always been a part of the writing process, usually at the stage of what writing teachers call 'prewriting.'[19] One could argue that the Internet has had the effect of bringing the primordial associative organization of texts back to the surface, of foregrounding association as 'finished writing' as opposed to 'prewriting.' During the age of the Gutenberg Galaxy, Western culture prized the hierarchies of sequence, linearity and logic as the hallmarks of an educated person because these were among the structural features of the printed book. Part of our discomfort with the Internet might stem from the fact that it does not appear and behave like a hierarchically ordered printed book.

Given its associational nature, perhaps it would be more useful to not think of the Internet as a writing space at all? Indeed, there are some thoughtful observers who would argue that comparing the Internet to a book or any other kind of writing space is the wrong analogy. For these observers, the Internet looks more like the 'cabinets of curiosity' that were fashionable in Europe from the 16th to the 18th centuries. These cabinets contained odd objects collected together, in a seemingly disorganized fashion, but in fact were linked together via association rather than taxonomic logic.[20] Cabinets of curiosity, as with many other artistic forms, work via analogy, meaning the ability 'to see coordination across separation...to couple data that is not effectively or invariably coupled by causal laws.' Stated another way, thinking via analogy means having the ability to understand similarity in the midst of apparent difference. For the art historian Barbara Maria Stafford, curiosity cabinets 'embody with great power and clarity the central idea of the analogical world view, namely that all physical phenomena...can be cross-referenced, linked in reconciling explanation by the informed imagination.'[21] When

DOI: 10.1057/9781137460950.0005

placed in the context of deep history, perhaps it makes more sense to view the Internet not as a writing space but as a globe-spanning analogical curiosity cabinet, a collection of curiosities that are cross-referenced and associatively linked, which would suggest a very different metaphor of the mind.[22] In privileging linearity over association, Western book-centric culture has either denigrated or ignored the many ways the mind works via analogy and associative connections. If the Internet is a representational space that enforces association, then it is simply mirroring our underappreciated associative brain.

The associative mind might appear to be an evolutionary step back from the logical textual mind enforced by the Book. But thinking of the mind as like a writing space may conceal from us its real configuration, perhaps the mind is not a book but rather a work of art. Conceiving of it in this fashion allows us to better understand 'the connective aspects of cognition' not as a disease, not as a symptom of a 'stupid brain' but as a more realistic reflection of how the brain has always worked. Yes, the brain is linear and logical and capable of deep reading. But the brain is also analogical and associative, capable of making connections between disparate objects and data points, and indeed has long been doing so. The Internet has not dulled our minds but has instead unleashed this pre-existing, if undervalued, portion of our cognitive architecture.

As I write this essay, researchers at UCLA, publishing in the Proceedings of the National Academy of Sciences, have proposed that the brain does not operate like a top-down hierarchical system, a view that neurologists have held for decades. Rather, these researchers contend that 'the brain appears to be a vastly interconnected network much like the Internet.'[23] This is still preliminary research, and would need to be replicated and confirmed, but there are ironic implications here: of an Internet-like brain structure that is being rewired by the Internet.

Notes

1   Nicholas Carr, 'Is Google Making Us Stupid?' *The Atlantic*, July/August 2008, http://www.theatlantic.com/magazine/archive/2008/07/is-google-making-us-stupid/6868/ See also Nicholas Carr, *The Shallows: What the Internet is Doing to Our Brains* (W.W. Norton and Co., 2010).

2   Carr, 'Is Google Making Us Stupid?'

3   Richard Foreman, 'The Pancake People, or, "The Gods are Pounding my Head,"' *Edge, The Third Culture*, http://www.edge.org/3rd_culture/foreman05/foreman05_index.html

4   Foreman, 'The Pancake People.'

5   Carr, *The Shallows*, 47. 'Though we are rarely conscious of the fact, many of the routines of our lives follow paths laid down by technologies that came into use long before we were born. It's an overstatement to say that technology progresses autonomously – our adoption and use of tools are heavily influenced by economic, political, and demographic considerations--but it isn't an overstatement to say that progress has its own logic, which is not always consistent with the intentions or wishes of the toolmakers and tool users. Sometimes our tools do what we tell them to. Other times, we adapt ourselves to our tools requirements.'

6   Carr, *The Shallows*, 51.

7   N. Katherine Hayles, *How We Think: Digital Media and Contemporary Technogenesis* (Chicago: University of Chicago Press, 2012), 10. 'The proposition that humans coevolved with the development and transport of tools is not considered especially controversial among paleoanthropologists. For example, the view that bipedalism coevolved with tool manufacture and transport is widely accepted. Walking on two legs freed the hands, and the resulting facility with tools bestowed such strong adaptive advantage that the development of bipedalism was further accelerated.'

8   Hayles, *How We Think*, 10.

9   Ibid., 11.

10 Lambros Malafouris, *How Things Shape the Mind: A Theory of Material Engagement* (Cambridge, MA: MIT Press, 2013), 247.

11 Stanislas Dehaene, *Reading in the Brain: The Science and Evolution of a Human Invention* (New York: Viking, 2009), 302.

12 Dehaene, *Reading in the Brain*, 7.

13 Ibid., 205.

14 Ibid., 6-7.

15 Plato, *The Phaedrus*, http://www.units.muohio.edu/technologyandhumanities/plato.htm

16 Matthew G. Kirschenbaum, 'The Remaking of Reading: Data Mining and the Digital Humanities,' *NSF Symposium on Next Generation of Data Mining and Cyber-Enabled Discovery for Innovation*, October 11, 2007, 1–2. http://www.csee.umbc.edu/~hillol/NGDM07/abstracts/talks/MKirschenbaum.pdf

17 Kirschenbaum, 'The Remaking of Reading,' 2.

18 J. David Bolter, *Writing Space: Computers, Hypertext, and the Remediation of Print*, second edition (Mahwah, NJ: Lawrence Erlbaum Associates, 2001), 13.

19 Bolter, *Writing Space*, 33.

20   This is the observation of Horst Bredekamp, in *The Lure of Antiquity and the Cult of the Machine: The Kunstkammer and the Evolution of Nature, Art and Technology* (Princeton: Markus Wiener Publishers, 1995), esp. 113.

21   Barbara Maria Stafford, *Visual Analogy: Consciousness as the Art of Connecting* (Cambridge, MA: MIT Press, 1999), 169.

22   Stafford, *Visual Analogy*, 139, 138 Many of the assumptions underlying the work in cognitive science assumes a linear logical, text-like brain. But Stafford wonders 'why does [cognitive science] look primarily to text-based fields, rather than the imaging arts, for insight on how cognition actually works?' 'Cabinets of curiosities, Piranesi etchings, cubist collages, dada-inspired boxes, even the Netscape browser or Macintosh's mosaic toolbar [she was writing in 1999] all provide information about some connective aspect of cognition that are not well captured by the scientific approaches currently adopted' (emphasis mine).

23   Jason Palmer, 'Brain works more like internet than "top down" company,' *BBC News*, August 10, 2010, http://www.bbc.co.uk/news/science-environment-10925841

DOI: 10.1057/9781137460950.0005

3
Query

Abstract: *This chapter presents a future scenario where the Internet develops into an autonomous 'brain.' That brain could emerge as one of four types: (1) a globe-spanning encyclopedia, an otherwise inert collection of all of the world's information that responds to our queries, (2) a network of the 'best brains,' allowing us to query a topic or problem and solving it through a kind of collective human intelligence, (3) a system driven by algorithms acting upon mountains of data that feeds back ideas and suggestions to us, the system being designed to 'nudge' us into action, and (4) a fully conscious and autonomous intelligence, possibly in competition with the biological brain.*

Keywords: algorithms; brain; cognition; data; future; human; information; intelligence; internet; networks

Staley, David J. *Brain, Mind and Internet: A Deep History and Future.* Basingstoke: Palgrave Macmillan, 2014. DOI: 10.1057/9781137460950.0006.

Nick Carr quotes, with some alarm, the musings of Google's founders, Sergey Brin and Larry Page, who 'speak frequently of their desire to turn their search engine into an artificial intelligence, a HAL-like machine that might be connected directly to our brains.' Carr quotes Page as saying 'The ultimate search engine is something as smart as people – or smarter,' and quotes Brin as suggesting 'Certainly if you had all the world's information directly attached to your brain, or an artificial brain that was smarter than your brain, you'd be better off.' Carr finds this vision of a supplemented brain unsettling: '[In this view] the brain is just an outdated computer that needs a faster processor and a bigger hard drive.'[1] Carr is concerned, I think, with (1) the idea of a competing brain to the human brain and (2) the seemingly unnatural connection between the two. Is the Internet in fact developing as a brain? If so, in what ways will we connect, and with what effects will we interact with, that brain?

Before we understand the meaning of the Internet as a 'Global Brain,' it would be useful to draw distinctions between four different connotations of the term, and how the Internet either already is or will one day develop into one of these 'brains.' 'Global Brain' has meant, for some, little more than a global encyclopedia, an otherwise inert collection of all of the world's information that responds to our queries. Some talk of a global brain as a network of the best brains, querying a topic or problem and solving it through a kind of 'collective intelligence.' A Global Brain might also refer to a system that is driven by algorithms acting upon mountains of data that feeds back ideas and suggestions to us, the system being designed to 'nudge' us into action. In effect, our queries to such an 'autonomous' brain would be indirect, seeing as they come from algorithms of our own invention. Finally, a global brain might be conceived and developed to be conscious and fully autonomous, perhaps in competition with the biological brain.

The idea of a Global Brain long predates Brin and Google.[2] On the eve of the Second World War, H.G. Wells imagined a global organization that would accumulate the knowledge of the world's best minds in one repository that would be easily and readily accessible to the entire population. As he was wont to do, Wells was imagining this as a secretariat of the brightest minds – an intellectual counterpoint to a world government – a kind of social network only open to the very smart, who would organize, direct and distribute knowledge through a 'world encyclopedia.'[3]

Wells maintained that this new organization of knowledge was befitting of the modern world. Indeed, he saw universities in particular as

DOI: 10.1057/9781137460950.0006

outdated relics of the medieval world. In referring to his vision as a World Brain, Wells was anticipating a way of extending cognition, of adding another layer to the brain, what he termed a 'new all-human cerebrum.'[4] Encyclopedism was hardly Wells' invention. Indeed, the desire to accumulate all of the knowledge of the world in an easy-to-access form is an ancient impulse. Thus, when Wells envisioned his Encyclopedia-cum-Global Brain as having 'the concentration of a craniate animal and the diffused vitality of an amoeba,' he was describing a long-standing desire to combine the contents of the world's libraries within a device as portable as a book. That his description sounds very much like the distributed network that is the Internet should not escape our notice.

Shortly after the Second World War, Vannevar Bush expressed similar desires as Wells. Given the explosion of knowledge and information, especially of scientific information, Bush noted that academic specialization was a necessary information management tool. But what was also needed was some way to bridge those disparate disciplines and the knowledge they held. Because scientists were producing so many nuggets of knowledge, it was difficult to navigate through that labyrinth of information to find what might be useful, especially to the nonspecialist. In his oft-evoked article 'As We May Think,' Bush described a system – again, using the technology of his day – that automatically records notable activities in a laboratory, that stores every article and scientific paper written, every conference proceeding, and organizes this information through a system of 'associative indexing' such that any working scientist would be able to access that information and, importantly, be able to quickly apprehend its meaning.

Bush called his visionary device a 'memex.'

> Consider a future device for individual use, which is a sort of mechanized private file and library. It needs a name, and, to coin one at random, 'memex' will do. A memex is a device in which an individual stores all his books, records, and communications, and which is mechanized so that it may be consulted with exceeding speed and flexibility. It is an enlarged *intimate supplement to his memory.*
>
> It consists of a desk, and while it can presumably be operated from a distance, it is primarily the piece of furniture at which he works. On the top are slanting translucent screens, on which material can be projected for convenient reading. There is a keyboard, and sets of buttons and levers. Otherwise it looks like an ordinary desk.[5] (emphasis mine)

DOI: 10.1057/9781137460950.0006

While Bush did not offer a visual representation of his memex in the article, one can easily conjure a vision of what Bush had in mind. The cynic would exclaim that Bush had described the modern office cubical, but I would prefer to think of the memex as an updated version of St Jerome in his study, a mechanized private library, only instead of being surrounded by books, rulers, globes, and other tools of the (ancient) scholar, 'Vannevar Bush at his memex' is surrounded with screens, buttons, and levers granting him access to the world's knowledge. While he did not use the term 'world brain', Bush was thinking along similar lines as Wells, in that the memex would serve to extend human memory in external form. More to the point, sitting at his memex allowed the scientist to off-load repetitive forms of thinking, allowing him to concentrate on higher-order thinking.[6]

Often when it is evoked, Bush's memex is understood as the forerunner of hypertext or of the associative linking of information such that one finds flitting across the Internet.[7] What I find more prescient about Bush's vision was the idea of the memex as an extension of the mind, as part of the architecture of human cognition. The memex was to solve the problem of too much information that needs to be stored, managed, and easily accessed, and that cannot be contained in the brain of just one person. Like St Jerome, the user of a memex cannot effectively engage in mental activity without being enveloped by his cognitive prosthesis, the 'study' replaced by the 'memex'.

In both cases, the 'brain' in question is more like a storehouse of information, a library or archive, inert until it is activated by a querying (human) brain. In Wells' imagination, this 'all-human cerebrum' was not an autonomous brain. That is, the World Brain does not initiate thought but instead offers up knowledge in response to our queries. In effect, the dream of a World Brain (World Encyclopedia) has been fulfilled with the Internet. Query any topic you like, and you receive an immediate response, as our Doonesbury character does in his class. With the Internet, we have off-loaded some of our cognition, we have supplanted our limited memories onto a vast World Encyclopedia, as immediately accessible to anyone carrying a portable memex.

Wells' World Brain or Bush's memex were both intended to store and distribute symbolic information, text, scientific papers, and the like. The Internet also connects individual brains, and there are those who would suggest that such linking of the world's best minds is also a way we extend cognition. The World Brain in Wells' imagination 'would be in continual correspondence with every competent discussion, every survey, every statistical bureau in the world,' and he indeed referred to his World Encyclopedia of experts as a network. Again, we are witness today to such a networking of great human minds, a network that itself is extending our collective cognitive abilities. Caroline Wagner describes the networking of scientists across the globe today as a 'new invisible college.'[8] Michael Nielsen would go even further in his assessment, arguing that given the amount of specialized information produced today, and given our ever-growing amount of Big Data, that such 'networked science' is the only way we will be able to do science in the future, since no one individual would be personally able to make all of the connections or to see new patterns in the ever-expanding amount of scientific information.[9] Networking lots of brains together facilitates and eases such connection-making. Nielsen describes the work of the Polymath Project, started by mathematician Tim Glowers who posited a complex mathematical problem, then invited the world's mathematicians to work collaboratively to solve it. The Internet fosters such massive collaboration, and the effects can be quite stunning: Glower's problem was solved in 37 days, with Glower describing the networked, collaborative problem solving approach as analogous to traditional research 'as driving is to pushing a car.'[10] Nielsen does not go so far as to call this network of minds a 'brain,' however, only a 'short-term working memory,' for indeed the kind of cognition occurring across global electronic networks is not comparable to the processes of the brain, however metaphorically suggestive that might be.[11]

Whether or not one wants to call this electronic network of experts a 'brain,' it is clear that working collaboratively with others over electronic networks amplifies the cognitive abilities of any one member of the network. In effect, 'networked science' and the 'new invisible college' represent efforts to off-load some cognition onto the larger electronic external symbolic storage system. In this case, that system is actual minds, not inert symbols. I wonder: how many other kinds of cognition will we be similarly enhanced and amplified by working with this 'new all-human cerebrum?'[12]

Nielson observes that as the store of data at our disposal grows ever larger, 'in many ways we are not so much knowledge-limited as we are question limited.'[13] That is, 'we're limited by our ability to ask the most ingenious and outrageous and creative questions' because we know that attempting to answer such questions would be impractical: either we lack sufficient data or, given a superabundance of data, we lacked the ability – as individuals – to comb through all of that data to discover meaningful patterns. It would take an army of brains working across the entire corpus to make new discoveries. With more data being produced, with more data shared widely across the electronic nervous system, and with more brains working together to apprehend the meaning of that data, we now have the conditions to ask a whole new set of questions. The scale of our queries – our ability to ask questions, that most fundamental of human cognitive abilities – is being enhanced by our participation in global networks.

The existence of a network of intelligent brains is hardly new, of course: the 'new invisible college' is in many ways an updated version of its 18th-century counterpart, helped along by electronic networks. The Republic of Letters connected the best minds of Europe with that analog networking technology. What is different here, I think, is our reconceptualization of the value of such networks, and our growing understanding that these networks have effects on cognition, on our ability to solve wicked problems. It suggests that interdisciplinarity cannot be contained inside the mind of one person but is a feature of such networked intelligence. The neuroscientist and entrepreneur Jeffrey Stibel wonders if having access to such a network of brains will change the way we think about human intelligence. 'Why value IQ,' he asks, 'when relationships are more powerful? After all, who is smarter when you have mental access to [all the world's] information – the person who knows all there is to know about quantum mechanics but nothing about nonlinear geometry, or the person who has close friends in both fields and just enough knowledge *to ask the right questions?*'[14] (emphasis mine). Stibel seems to be suggesting that query – being able to ask questions, especially questions that extend beyond intellectual silos – will be the mark of intelligence in a world where our brains can so easily access the World Encyclopedia. Whether or not this networking of brains reflects a new kind of artificial, globe-spanning brain seems to me less important than the larger notion that, through external electronic means, humans are expanding our cognitive capacities, very much in keeping with our evolutionary sense of technogenesis: that our brains emit new systems of external cognitive enhancement as a way to amplify the capacity of our brains, and in so doing expand the architecture of the mind.

The kind of online collaboration imagined here is not a typical crowd-sourced social network, where any ill-considered opinion (or rant) is offered. Wells' World Encyclopedia was to harness the collective wisdom of the best minds; Bush imagined a storehouse of peer-reviewed scientific information. Querying the Internet, as we all know, does not always yield 'the best that is thought and known in the world.' There are obviously many outstanding sources available to us on the Internet. However, there are toxic sources as well. Will the information about greenhouse gasses that our cartoon student accessed be subsumed under mountains of silly articles about reality-show celebrities? Just as our bodies require nourishment and are harmed when we ingest junk food, junk information has analogous effects on the mind. Similar arguments were made about television, of course. But this should not prevent us from raising these questions again regarding the Internet, and especially consider the potential for the rapid expansion of new forms of junk information.

While we might question the value of much of what passes across the Internet, there are other outstanding sources of information for us to query. Why search Google when a query to Wolfram Alpha would yield much more reliable results? Google Books has its sights set on becoming the *de facto* Global Library, to the consternation of many concerned about a for-profit entity with that much control over the world's storehouse of knowledge. But the Digital Public Library of America (DPLA)– a nonprofit intuitive championed by academics – aims to bring together 'the riches of America's libraries, archives, and museums, and [make] them freely available to the world. It strives to contain the full breadth of human expression, from the written word, to works of art and culture, to records of America's heritage, to the efforts and data of science.'[15] Wells' vision of a World Brain where 'the whole human memory [is] made accessible to every individual' is well upon us.

To reiterate, the Global Brain we are referring to here is not autonomous, engaging in self-generated cognition. When we access a text from the DPLA, we are activating the Global Brain by our query. We all watched in awe as IBMs Watson easily defeated the best human *Jeopardy* players. But what we witnessed was not a conscious brain besting human brains, but rather a very good but ultimately glorified search engine. Watson is a Global Brain in the Wellsian sense: a repository of information that responds to our (or Alex Trebek's) queries. We are very close to the day when we will be able to carry a device in our pockets that we will query with our voices and gain access to the search capabilities of IBMs massive parallel-processing computer. That figurative marriage between Siri and Watson will no doubt extend our cognition in ways Wells never

DOI: 10.1057/9781137460950.0006

imagined, and with a portability Bush – confined to his desk – could not envision.[16]

The networking of brains reflects an external symbolic storage system more 'animated' than the inert Encyclopedia model. That is, what is being networked in such an invisible college are many conscious, autonomous brains that need not be activated by query alone. What happens when that external Encyclopedia anticipates our queries, or indeed makes suggestions without our making a direct query? What happens when our external memory systems begin to query us? What kind of cognition would this represent?

To take but one example, Clive Thompson observes that we are just now starting to 'embed prospective memory in the world around us.' By this, he means that, in contrast to retrospective memory – information from the past as might be contained in an encyclopedia or library – prospective memory involves remembering to do something in the future. Current tools are wedding calendars and to-do lists with GPS systems to embed reminders in our physical environment. 'We've long wrangled with to-do lists to remind us *what* we need to do,' observes Thompson, 'and calendars to remind us *when* we need to do it. But what we also need is to be reminded *where* to do things – a nudge that occurs not just in time but in space.'[17] As I write this sentence, my phone has just vibrated, reminding me 15 minutes before the event that I have a meeting at 11:30. Thompson is describing a system that would embed such a reminder not only at a specific time but also in a specific location, that location triggering an association to remember such-and-such an event. A prospective memory system 'could let you *embed your intent* inside your house or office – reminders that call out at the right time as you pass by [as] when I walk past the fridge [and] "take out the meat that I need to defrost for tonight"' pings my device[18] (emphasis mine). My query here is indirect: I don't speak aloud or otherwise directly address the external memory system. That system responds to my indirect query, in that I have 'embedded my intent' to be reminded to take out the meat, but such a response to my query comes only as I carry out my other (unconscious) activities. Unless I actually walk by the refrigerator, the query goes unanswered.

Embedding intent in the larger environment represents a manifestation of the extended mind hypothesis, when 'the environment' is responsive to our queries, and aids us in cognition. The above refrigerator scenario is sometimes categorized under the Internet of Things

DOI: 10.1057/9781137460950.0006

scenario: the idea that all of our devices, even the smallest and seemingly un-information-laden, will nevertheless produce information that will be networked with other such devices. In a typical home, for example, my refrigerator will produce information about its contents, and that information will be shared with other devices in my house, such as my medicine cabinet, which is monitoring my prescriptions.[19] The vision for the Internet of Things is that the refrigerator and medicine cabinet and prescription bottles will be 'communicating' with each other such that, when I return from the grocery and place a pint of ice cream in the freezer, my medicine cabinet will be alerted. Were I to have a statin prescription, my cabinet would 'know' that I have high cholesterol, and ice cream represents an incongruity. This system, then, would ping me in some fashion, and say to me something like 'Do you really need that ice cream? You'd better watch your intake, as this could adversely affect your cholesterol levels.' It is possible in such a scenario that this incongruity could be 'reported' to my doctor, this information added to my patient profile, triggering a warning from my doctor (or perhaps his electronic surrogate) to be mindful of my cholesterol.[20] In effect, this Internet of Things will be acting as my electronic Jiminy Cricket, a 'conscience' or 'voice in my head' steering me in the right direction. I would be off-loading some form of cognition onto the larger system, an extension of my mind onto the larger electronic environment.

When I make decisions and take actions, like purchasing ice cream, I am in effect querying the electronic system, even though I am not phrasing it in the form of a direct question. Presumably, I have consented to having my dietary choices so monitored, giving my tacit approval and 'embedding my intent' in the environment around me. The question becomes: what if someone else's intent is embedded in my environment? What if that Internet of Things that surrounds me is of someone else's design that responds to my indirect queries? It is perhaps not coincidental that Thompson used the term 'nudge' to describe his prospective memory system, the term resonating with the title of Cass Sunstein and Richard Thaler's book about 'choice architecture.' Choice architecture involves 'organizing the context in which people make decisions.'[21] A context might be designed in order to provide a 'nudge' to a user, which Sunstein and Thaler define as 'any aspect of the choice architecture that alters people's behavior in a predictable way without forbidding any options or significantly changing their economic incentives.'[22] In my ice cream scenario above, one could argue that that context has been designed to

DOI: 10.1057/9781137460950.0006

provide me a nudge, 'shaming' me into not eating ice cream. Of course, I could just as easily ignore that electronic voice in my head and eat the ice cream anyway, that option not being expressly forbidden. But the electronic voice nudges me to make the right dietary choices, even if I lack the willpower to make such a choice myself. In this nudge scenario, however, it is the choice architect's intent that has been embedded in my environment, not mine. Perhaps this is the root of Carr's concern? That the 'intent of others' embedded in my environment is Google's or some other multinational or someone who has other interests than mine, with control over my queries? Perhaps the concern is that the 'system' itself develops a consciousness that is outside of any (human) control?

The choice architect in the above instance is the writer of the code, the designer of the algorithms that connect my devices together. Like a Great Watchmaker, the code writer develops the algorithms that act upon the information generated and networked across my devices. Indeed, the Internet of Things continues to grow, as every choice I make, every action I conduct leaves digital traces, data and information that are being collected and archived. Through both our conscious, deliberate choices and through scores of involuntary actions, we produce mountains of data, so much data, in fact, that scientists today talk about 'Big Data.' Computer scientists maintain that we are now in a position to not only collect and store that data but that we are also able to perform computations on that data, seeking out patterns that will allow us to better predict human behavior.[23] Choice architects will no doubt find these patterns of human behavior useful for designing systems to nudge our actions.

How might any one person query, let alone apprehend meaning from, these very large data sets? Coders and developers are now developing tools and algorithms that will operate on data and information directly to discern pattern and meaning.[24] Brian David Johnson refers to the ever-increasing amounts of data produced and captured and the algorithms that act upon that data as the 'secret life of data.' A futurist at Intel, Johnson imagines that in the near future 'data will have a life of its own.'[25] That data will come to life because 'we will have algorithms talking to algorithms, machines talking to machines, machines talking to algorithms, sensors and cameras gathering data, and computational power crunching through that data, then handing it off to more algorithms and machines. It will be a rich and secret life separate from us,'[26] in that those operations will go on automatically without our conscious knowledge and tracking. In the scenario I am imagining, we will indeed

DOI: 10.1057/9781137460950.0006

interact with that 'secret life' both through direct query but also when that secret life nudges us to make specific choices. That data will not be the inert data of the archives or the World Encyclopedia: data will be alert and active, and perhaps even proactive.

The provost at Austin Peay State University, Tristan Denley, demonstrates how algorithms acting upon Big Data builds a choice architecture for students. Denley wondered if he could analyze the grade data the University produced, the only use of which was for transcripts. Denley instead actively analyzed all of this inert data to see if there were behavioral patterns that could be used to predict student outcomes. That is, he wondered if he could correlate student performance in a given class by comparing the performances of similar students. He 'wondered if he could nudge students to choose courses that would lead to a degree, if they knew in advance how well they might perform.'[27] Students would often take classes without any real sense of their potential for success, would fail at these classes, and therefore slowed their progress toward a degree. Having better information about their potential for success, Denley reasons, would nudge students to choose classes more wisely. At present, students are advised in their choices by advisors 'influenced by the advice they have given in the past.'[28] During testing of Denley's software, a student was 'advised' by the algorithm to take Arabic, even though the human advisor wondered why the choice suggested was not Spanish. When pressed, the advisor said that he wondered about this 'Because most of the time I recommend Spanish. Most people take Spanish.' One conclusion we may draw is that an algorithm acting on very large data sets might provide more reliable advice than a human being acting only on 'experience.'[29]

I provide this example only to suggest where choice architecture and embedded intent might be headed: given that our society is producing more and more data that is open to capture and analysis, and given the proposition that analyzing Big Data reveals patterns in those data that can help us predict a variety of outcomes, that we will see such a combination of 'algorithm acting on data' informing more of our choices. At present we similarly receive recommendations about movies or books from Netflix or Amazon. These automated recommendations derive from algorithms acting upon data and finding patterns that might overwhelm the attention of an individual human brain. We should anticipate going forward that such electronic aids-to-cognition will expand to nudge us into a host of other choices.

DOI: 10.1057/9781137460950.0006

'Algorithms acting upon information' strikes me as the kind of Global Brain Sergey Brin is discussing, and that Carr finds so alarming. Is such an autonomous external system a 'brain?' Or is it more like a thermostat, as cyberneticists have long held? While the secret life of data appears to be humming along without our participation or consent, we must understand that, in this formulation, the 'brain' in question is still a human brain, or at least the human brain that has designed the algorithm. But, clearly, as long as these algorithms are embedded in our environment and as long as we are influenced and nudged by the results of these 'algorithms acting upon data,' then at least some of our cognition is being 'outsourced.' When Glowers posted that complex mathematical problem to the collective intelligence of other mathematicians, he was in effect 'querying' a larger network, and he received his response. Will our electronic networks, driven by algorithms, reach the stage of development such that they, too, will be able to answer more sophisticated queries, beyond information retrieval or simple pattern recognition, to be able to offer answers to wicked problems? At that stage, our algorithms will have moved beyond pattern recognition to being able to act – independently of our intent – upon those patterns.

The Internet is a global nervous system, in the words of Al Gore: 'the simultaneous deployment of the Internet and ubiquitous computing power have created a planet-wide extension of the human nervous system that transmits information, thoughts, and feelings to and from billions of people at the speed of light.'[30] At one level, the Internet does seem to mimic our neurological system: electrical pulses coursing along networks. The Internet is fine as long as it remains merely a 'nervous system;' it is when we use the term 'brain' to describe this network that we start to get concerned, for 'brain' implies not just electric impulses, but consciousness and thinking, and an autonomy from humanity, 'a very complex organism that often follows its own urges.'[31] In none of the examples above are we suggesting that the Internet is an actual brain. That is not to suggest, however, that there are those who see the Internet as an emerging brain, a vastly superior brain to our biologically limited brain.

In one telling, the structure of the Internet looks very brain-like. The nodes of computers, servers, other human brains might be compared to neurons, the branching networks connecting those neurons appearing very much like dendrites and axons. What if these comparisons were not merely metaphorical but actual? Artificial intelligence (AI) researchers have for over half a century sought to replicate the brain in the computer,

DOI: 10.1057/9781137460950.0006

and have scored notable successes, as when the IBM computer Deep Blue defeated Garry Kasparov in a series of chess matches, or when the aforementioned Watson easily defeated the best *Jeopardy* champions. In one narrative, these are chapters in a triumphant story of our ability to recreate the human brain via artificial means. Perhaps what has changed is the way we will think about this brain: from being contained in a single object to being distributed across a vast network, the ultimate act of parallel processing. Rather than sitting in one device, the artificial brain will extend across the entire planet, the operations of the human brain copied, enhanced, and transcended via our globe-spanning network.[32]

Ray Kurzweil is one futurist who believes that these tools will allow us to – very soon – replicate the brain *in silico*. Specifically, he believes that we will create an artificial neocortex, based on his 'pattern recognition theory of mind,' which he maintains 'describes the basic algorithm of the neocortex.'[33] In Kurzweil's formulation, the neocortex is a vast store of patterns, and our brains contain 'pattern recognition modules' that allow us to identify and act upon those patterns. Eventually, believes Kurzweil, we will develop algorithms that will similarly store, recognize, and act upon patterns. Rather than simply nudging us into action, our 'algorithms acting upon data' will decide for themselves how and when to act. At that stage, we would have indeed replicated the neocortex in digital form, and our planet-wide human nervous system would have developed into a human brain.

Kurweil refers to that moment when the artificial neocortex starts behaving like a human brain as The Singularity, that moment when 'information-based technologies will encompass all human knowledge and proficiency, ultimately including the pattern-recognition powers, problem-solving skills, and emotion and moral intelligence of the human brain itself.'[34] Because of what he terms the Law of Accelerating Returns – a point to which we will return in the next chapter on 'Limits' – 'the nonbiological portion of our intelligence will be trillions of trillions of times more powerful than unaided human intelligence.'[35] Stated another way, the electronic external symbolic storage system will itself become 'intelligent,' and not merely an archive or storehouse of symbols. Because of the vast computing power and parallel processing enabled by the Internet, the Library and the Encyclopedia will have become conscious and sentient.

Kuzweil's vision is based on the understanding that computing power will continue to grow exponentially. As long as we assume that the human brain and its vast capacities are based solely on what happens in the neocortex, then it is easy to extrapolate a future where its pattern-recognition functions will be replicated across the Internet. If only the human brain worked in solely that manner: even in the wake of Deep Blue's historic victory, many AI researchers acknowledge that the idea of replicating or surpassing the brain *in silico* remains a daunting challenge. Computer success at chess or Jeopardy is one thing; Deep Blue could not play checkers, however, and Watson was not able to reflect upon the meaning of its victory, cognitive tasks that humans perform with relative ease. Throwing a ball or skipping rope remain simple cognitive tasks that are nevertheless beyond the ability of a computer or computer network, in part because it lacks a body. Indeed, a large part of human intelligence comes from the fact that we possess bodies that move through space.[36] Replicating the neocortex would certainly allow us to offload many cognitive tasks, but would still leave us some distance from the full capacity of the human brain.

Even if we are unable to replicate the human brain, one could argue that the Internet is developing into something more than a nervous system. Jeffrey Steibel, for one, thinks the internet is a brain, just not a human brain. Steibel observes that the structure of the Internet appears similar to the structure of the human brain, but his view diverges from Kurweil's in understanding that the Internet-as-brain is not coterminous with the human brain.

> When the Wright brothers first flew… their intent was not to create a bird. To be sure, some innovators thought that building a 'bird' was the road to flight, but it was not. The Wright brothers harnessed the laws of flight, and not the body of a duck or a blue jay… humanity was going to accomplish flight in its own way. And we did achieve flight on our own terms, without replicating what nature had done. The same can be said of Internet intelligence. It will no more look like a brain than an airplane looks like a bird. Nor is it going to act like a human being.[37]

An airplane is not a bird. Today's airplanes can do things birds cannot, in terms of speed, distance, height, the ability to carry passengers, and so on. That is, the airplane has extended far beyond merely mimicking birds. It should also be stated that birds continue to do things that airplanes cannot, in terms of maneuverability, self-replication, and so on. Similarly, the Internet-as-brain will not merely mimic the human brain. It will certainly perform cognitive functions, it might even assert its own will, its own urges beyond merely nudging us into action. But all of this cognitive activity need not be understood as mimicking the human brain: it will represent a new kind of brain. The Internet will be a brain the same way a bird or a dog or a cheetah has a brain. It will be a nonhuman brain.

Will that brain be better, more advanced than our human brain? That, again, depends on what we mean by 'better.' Is a calculator 'better' than me because it calculates faster than I do? Is Deep Blue 'more advanced' than my brain because it can outplay me at chess? I think that the real question is not whether the Internet-as-brain will become superior to us, and all of the science fiction-inspired dystopias that represents. Rather, the question will be whether the internet will become an autonomous intelligence and if so, how will we negotiate and interact with the alternative intelligence? The designer Donald Norman compares this to riding a horse. A horse has its own brain, its own intelligence, and yet a skilled rider forges a relationship with that autonomous intelligence. Learning to cooperate with that autonomous intelligence is how I see our relationship with the Internet going forward: we will develop a choreography that will organize the symbiotic relationship between these two complementary intelligences. When the Internet begins to query us, how will we respond?

Notes

1   Nicholas Carr, 'Is Google Making Us Stupid?' *The Atlantic*, July/August 2008, http://www.theatlantic.com/magazine/archive/2008/07/is-google-making-us-stupid/6868/; *The Shallows: What the Internet is Doing to our Brains* (New York: W.W. Norton, 2010).

2   Carr discusses the vision of a global library, and also the legal and technical issues of building one, in 'The Library of Utopia,' *MIT Technology Review*, April 25, 2012, http://www.technologyreview.com/featuredstory/427628/ the-library-of-utopia/ See also Peter Singer, 'Whither the dream of the universal library?' *The Guardian*, April 19, 2011, http://www.theguardian.com/ commentisfree/2011/apr/19/moral-imperative-create-universal-library On the long and problematic history of the idea of a Universal Library, see Ken Hillis, Michael Petit and Kylie Jarrett, *Google and the Culture of Search* (New York: Routledge, 2013), 77–104.

3   H.G. Wells, *World Brain* (Garden City, New York: Doubleday, Doran and Co., 1938), 69–70. 'A world Encyclopedia no longer presents itself to a modern

DOI: 10.1057/9781137460950.0006

imagination as a row of volumes printed and published once for all, but as a sort of mental clearing house for the mind, a depot where knowledge and ideas are received, sorted, summarized, digested, clarified and compared. It would be in continual correspondence with every competent discussion, every survey, every statistical bureau in the world. It would develop a directorate and a staff of men of its own type, specialized editors and summarists. They would be very important and distinguished men in the new world. This Encyclopedia organization need not be concentrated in one place; it might have the form of a network. It would centralize mentally but perhaps not physically. Quite possibly it might to a large extent be duplicated. It is its files and its conference rooms which would be the core of its being, the essential Encyclopedia. It would constitute the material beginning of a real World Brain.'

4   Wells, *World Brain*, 87. 'The whole human memory can be, and probably in a short time will be, made accessible to every individual. And what is also of very great importance becomes continually more frequent and unpredictable, is this, that photography affords now every facility for multiplying duplicates of this – which we may call? – this new all-human cerebrum. It need not be concentrated in any one single place. It need not be vulnerable as a human head or a human heart is vulnerable. It can be reproduced exactly and fully, in Peru, China, Iceland, Central Africa, or wherever else seems to afford an insurance against danger and interruption. It can have at once, the concentration of a craniate animal and the diffused vitality of an amoeba.'

5   Vannevar Bush, 'As We May Think,' *The Atlantic*, July 1, 1945, http://www.theatlantic.com/magazine/archive/1945/07/as-we-may-think/303881/

6   Bush, 'As We May Think.' '[Man] has built a civilization so complex that he needs to mechanize his records more fully if he is to push his experiment to its logical conclusion and not merely become bogged down part way there by overtaxing his limited memory. His excursions may be more enjoyable if he can reacquire the privilege of forgetting the manifold things he does not need to have immediately at hand, with some assurance that he can find them again if they prove important.'

7   Carr notes, for example, that 'With his memex, Bush anticipated both the personal computer and the hypermedia system of the World Wide Web,' especially drawing attention to Bush's concept of 'associative indexing.' See *The Shallows*, 169–170.

8   Caroline S. Wagner, *The New Invisible College: Science for Development* (Washington, DC: Brookings Institution Press, 2008), 2. 'Self-organizing networks that span the globe are the most notable feature of science today. These networks constitute an invisible college of researchers who collaborate not because they are told to but because they want to, who work together

not because they share a laboratory or even a discipline but because they can offer each other complementary insight, knowledge, or skills.'

9 Michael Nielsen, *Reinventing Discovery: The New Era of Networked Science* (Princeton: Princeton University Press, 2012), 91.

10 Nielsen, *Reinventing Discovery*, 2–3. Online collaboration of this sort is useful, Nielsen argues, because 'even the best mathematicians can learn a great deal from people with complementary knowledge, and be stimulated to consider ideas in directions they wouldn't have considered on their own,' echoing the observations of Wagner. 'Online tools create a shared space where this can happen,' notes Nielsen, 'a short-term collective working memory where ideas can be rapidly improved by many minds. These tools enable us to scale up creative conversation, so connections that would ordinarily require fortuitous serendipity instead happen as a matter of course. This speeds up the problem-solving process, and expands the range of problems that can be solved by the human mind.'

11 Nielson says 'A common approach to these questions is to suggest that online tools enable some sort of collective brain, with the people in the group playing the role of neurons. A greater intelligence then somehow emerges from the connections between these human neurons. While this metaphor is stimulating, it has many problems... Whatever our collective brain is doing, it seems likely to work according to very different principles than the brain inside our heads. Furthermore, we don't yet have a good understanding of how the human brain works, so the metaphor is in any case of limited use at best.' Nielsen, *Reinventing Discovery*, 18–19.

12 See especially the work of the MIT Center for Collective Intelligence, http://cci.mit.edu/

13 Nielsen, *Reinventing Discovery*, 103.

14 Jeffrey Stibel, *Wired for Thought: How the Brain is Shaping the Future of the Internet* (Boston: Harvard Business Press, 2009), ii–xiii.

15 The Digital Public Library of America, http://dp.la/info/

16 Marc Prensky lists a number of such cognitive enhancements when our brains are tethered to networked technologies, among which include:
Allowing us to 'do things in parallel' (85)
Turning all texts into verbal language, ensuring greater levels of literacy (87)
Translating words instantaneously, thereby eliminating language barriers (91–92)
Allowing us to know what others and thinking and feeling (100–101)
Collecting and analyzing more data (115)
'Reading' everything ever written (122–123)
Making faster decisions (131–132)
Analyzing risk faster and more prudently (133–134)
Connecting more ideas together (151–152)

DOI: 10.1057/9781137460950.0006

See Marc Prensky, *Brain Gain: Technology and the Quest for Digital Wisdom* (New York: Palgrave Macmillan, 2012).

17   Clive Thompson, 'A Sense of Place,' *Wired*, February 2013, 34.

18   Thompson, 'A Sense of Place,' 34.

19   On the idea of a kitchen with 'ambient intelligence,' see Donald Norman, *The Design of Future Things* (New York: Basic Books, 2007), 28–31.

20   On the implications of such networking for medicine and health care, see Eric Topol, *The Creative Destruction of Medicine: How the Digital Revolution Will Create Better Health Care* (New York: Basic Books, 2012).

21   Richard H. Thaler and Cass R. Sunstein, *Nudge: Improving Decisions About Health, Wealth, and Happiness* (New Haven: Yale University Press, 2008), 3.

22   Thaler and Sunstein, *Nudge*, 6.

23   Albert-Laszlo Barabasi observes that, because we are able to capture more and more of the data that we collectively produce every day, we will very soon be able to uncover the underlying patterns in human behavior such that we will be able to predict (and control?) that behavior. See *Bursts: The Hidden Pattern Behind Everything We Do* (New York: Dutton, 2010).

24   Nielson, *Reinventing Discovery*, 111. 'Rather than mining that knowledge in a piecemeal way, we'll be able to do automated inference on all of human knowledge, finding hidden connections ... What's gradually emerging is an online network of knowledge that's intended to be read by machines, not by humans. Those machines will find meaning in that network of knowledge, and help explain it to us.'

25   Brian David Johnson, 'The Secret Life of Data in the Year 2020,' *The Futurist*, July–August 2012, 21. 'With the massive amount of sensors we have littering our lives and landscapes, we'll have information spewing from everywhere. Our cars, our buildings, and even our bodies will expel an exhaust of data, information, and 1s and 0s at an incredible volume.'

26   Johnson, 'The Secret Life of Data,' 21.

27   Jeffrey J. Selingo, *College (Un)Bound: The Future of Higher Education and What It Means for Students* (Boston: New Harvest, 2013), 82.

28   Selingo, *College (Un)Bound*, 82.

29   Ibid., 82–83. 'One problem with human advisors is that their knowledge of an inch-long course catalog with hundreds or thousands of classes and dozens or perhaps even hundreds of majors is limited. They often know [only] their field best. When students need guidance on classes outside the major or even in other majors, advisors struggle and sometimes make bad recommendations.'

30   Al Gore, *The Future: Six Drivers of Global Change* (New York: Random House, 2013), 44.

31   Gore, *The Future*, 47.

DOI: 10.1057/9781137460950.0006

32 In 2002, Ben Goertzel wrote that 'The Internet as it is today is just a little baby. But it's on the verge of a fundamental transition. Today it's a distributed network of content and software, serving diverse people diverse functions. Soon enough it will be a self-organizing intelligent system, with its own high-level coherent patterns, serving not only as a mind but as a world inhabited by a diversity of digital life forms.' (ix) See *Creating Internet Intelligence: Wild Computing, Distributed Digital Consciousness, and the Emerging Global Brain* (New York: Kluwer Academic/Plenum Publishers, 2002).

33 Ray Kurzweil, *How to Create a Mind: The Secret of Human Thought Revealed* (New York: Viking, 2012), 5–6.

34 Ray Kurzweil, *The Singularity Is Near: When Humans Transcend Biology* (New York: Penguin, 2005), 8.

35 Kurzweil, *The Singularity*, 9.

36 An interesting meditation on this idea is George Lakoff and Mark Johnson, *Philosophy In The Flesh: The Embodied Mind And Its Challenge To Western Thought* (New York: Basic Books, 1999).

37 Stibel, *Wired for Thought*, xxiv.

DOI: 10.1057/9781137460950.0006

4
Interface

Abstract: *This chapter presents a scenario where the Internet and the brain couple such that they work together as two 'hemispheres.' Far from being overwhelmed or made vestigial by technology, our biological brain would work in partnership with our technological brain. The chapter explores the possibilities for not only direct implants, but also devices that convert information into sounds, smells, and movements, symbols arriving to us through all of our senses. As the logistics of access to information changes, the meaning of formal education will also be changed. How will we design formal education when we need not be tethered to physical centers of the symbolic storage system? How will we design formal education when our symbols are sonic and haptic, not just oral and textual?*

Keywords: brain; education; future; information; interface; internet; symbolic

Staley, David J. *Brain, Mind and Internet: A Deep History and Future*. Basingstoke: Palgrave Macmillan, 2014. DOI: 10.1057/9781137460950.0007.

I concluded Chapter 3 with a vision of the future where the human brain and an artificial brain work together like a horse and rider. The metaphor comes from the designer Donald Norman, who observes that:

> In order to cooperate usefully with our machines, we need to regard the interaction somewhat as we do interaction with animals. Although both humans and animals are intelligent, we are different species, with different understandings and different capabilities. Similarly, even the most intelligent machine is a different species, with its own set of strengths and weaknesses, its own set of understandings and capabilities. Sometimes we need to obey the animals or machines; sometimes they need to obey us.[1]

There is a complex boundary between those two states: the need to obey the machine and the need for the machine to obey our queries. Norman refers to the negotiation of that boundary, that interface between human brain and external system of symbolic storage, as symbiosis, the 'merger of two components, one human, one machine, where the mix is smooth and fruitful, the resulting collaboration exceeding what either is capable of alone. We need to understand how best to accomplish this interaction, how to make it so natural that training and skill are usually not required.'[2] I am struck by the terms Norman uses here: 'fruitful,' 'collaboration,' 'exceeding the capabilities of both.' The language here is one of cooperation with our technologies, not the opposition enforced by both technological determinists and instrumentalists. The design challenge for the future will be the interface between the human brain and the Internet-as-electronic-extension of that brain.

The post-Deep Blue narrative revolves around computers and humans working together to think together, or, in this case, play chess together. One lesson we learned from the Deep Blue-Kasparov competition is that computers are powerful aids to the human mind. Kasparov himself has said that he views computers as partners with humans to explore the intricacies of chess in ways not possible before. That is, powerful computers enable us to do some things faster or better. Such tools have, historically at least, extended – rather than supplanted – our cognitive capacity.

Michael Neilson reports about an online chess tournament which pitted 'hybrid teams' of humans and computers against each other. Human players teamed with computers and their database of opening and endgame moves. At times, the machines decided the moves, and at other times humans decided the moves.[3] One of the entrants was the Hydra computer chess program, which had regularly defeated even

DOI: 10.1057/9781137460950.0007

human grand masters, and indeed, in this tournament had easily defeated both human grandmasters and computers. But, interestingly, the hybrid teams of computers and humans easily defeated the Hydras. The reason was that the human players understood when to rely on machine intelligence and when to use their own judgment. Astoundingly, the tournament winner was the ZackS team, the hybrid team consisting of amateur players using relatively simple tools and software.[4] The implications here are that human intelligence alone could not defeat the computers, but that a computer alone also could not defeat the top computer. A human-computer team plays even better chess than any one brain – biological or mechanical – in isolation.

How will the symbiosis between the data-driven intelligence of the Internet collaborate with our biological brains? What will be the nature of that interface? Sergey Brin envisions 'the world's information directly attached to your brain.' I want to consider what this 'attachment' might look like, and how it will impact cognition. If we accept that the Internet represents some kind of 'brain' – whether inert and library-like or active and intentional – how will our biological brain interact with this electronic brain? What will the relationship be? Vannevar Bush described the memex as 'an intimate supplement to [the scientist's] memory.' I would like to understand the degree of intimacy that this interface between the biological and the electronic brain represents. Have such interfaces with our systems of external symbolic storage been similarly so intimate? What is the nature of the 'partnership' such brain/internet coupling represents?

In Chapter 3, I described a scenario where an electronic Internet of Things 'nudges' me to avoid eating ice cream. I suggested that an electronic 'voice' would caution me about what consuming the ice cream might do to my cholesterol levels. How will that voice come to me? It is obvious to observe that the devices that link us to the Internet are becoming smaller and more portable: Bush's memex can now be carried around in my pocket. Google Glass points in the direction of computer screens and interfaces sitting right in front of our eyes: that my electronic nudge might come in the form of a warning sign that appears in my peripheral vision. Others have posited contact lenses or other such devices that would mean that the screen would become indistinguishable from our field of vision, that the world we would see is both physical and digital. The short film 'Sight' by Eran May-raz and Daniel Lazo features a near-term future scenario where a character has devices either layered on a contact lens or somehow embedded in the eye that superimposes digital information over his physical surroundings.[5] The film is intended as an exploration of the 'gamification of life,' but what I find interesting is, first, the vision for how digital information will be seamlessly embedded into our surroundings, and how the interface between biological and electronic brain will be negotiated. The game player relies on just-in-time information to appear in front of his eyes, in this case, providing suggestions about what to eat, what to wear, what to say to the women he is trying to impress on a date. His electronic brain nudges him to make certain decisions; visual perception and electronic nudging becoming almost indistinguishable. What is also interesting about the film is the depiction of how our user, enthralled by the electronic information, appears vacant and in a trance when scanning the screen in front of his eyes: he is staring off someplace else, attending to his data, not his human companion. He is enjoying greater intimacy with his symbols, until he shifts his gaze back to his companion. When the distance between the screen and the eye has all but vanished, digital symbols and the real world will compete for our attention.[6]

The memex-in-our-pocket is about to become the memex-in-front-of-our-eyes. Wearable screens feel unsettling because that separation – the physical distance between our bodies and our external symbols – is gone, even if we have always harbored a kind of intimacy with our symbols. In the film 'Sight,' the connection between electronic information and brain is more direct, seemingly without that space. When we wear Google Glass or have lenses placed over our eyes, what is missing is the distance, the space between eye and book or eye and screen or eye and painting. While we might be intimate with a book ('getting lost in a book') there remains a physical distance between cognitive object and brain/body that is comfortable, even if there is a conceptual coupling that occurs when we read a book or view a painting. What will it mean for us to have the World Encyclopedia, not at our fingertips, but at a glance of our eye?[7]

The neuroscientist Robert Sapolsky observes that 'we humans are pretty impressive when it comes to being able to extract information, to discern patterns from lots of little itsy-bitsy data points,' like a musician reading sheet music or a scientist making a judgment. He sides with others who have marveled at the rising tide of data and information humans (through our technologies) have been producing, but laments our ability to make sense of it all. Repeating the concerns of scientists such as Vannevar Bush, Sapolsky observes that 'far too often the technologies

DOI: 10.1057/9781137460950.0007

have out-stripped our abilities to be insightful. We have some crutches – computer graphics allow us to display a three-dimensional scatter plot, rotate it, change it over time. But still, we barely hold on.'[8] It is interesting that he uses the term crutches, implying a temporary aid that we employ when because we are infirm.

Sapolsky notes that we have developed data visualization tools to aid us in the process of making sense of our avalanche of information. 3-D graphics allow us to plot data (numerical and, increasingly, textual data) such that we can draw insights from it. 'Looking at pictures,' at least in Gutenbergian Western culture, has been associated either with mere aesthetics or with a lesser form of knowing (note Pope Gregory's notion that cathedrals are for illiterates). After about the 1990s, and especially after the publication of Edward Tufte's *Visual Display of Quantitative Information*, did the idea that visualization and visual information could be composed with the same thoughtfulness and rigor as a prose composition did the idea gain traction that at least some visual displays could be more than just 'pretty pictures,' and that these might allows us to more quickly and efficiently apprehend meaning from large amounts of information. Martin Wattenburg's 'Map of the Market'[9] is one such example. Rather than attempting to read through tables of numbers, stock market information can be easily grasped in a visualization that quickly identifies different sectors, the market capitalization of different companies in that sector, and how those company's stocks are performing at any given moment. Information designers speak of the 'dimensions of information' such well-crafted visualizations can easily encompass in a relatively small space. Building displays in Sapolskian three-dimensional space adds to the amount of information that can be contained within.

Such 3-D environments of data, anticipates Sapolsky, will be the way that the human brain will be able to cope with increasing amounts of data. 'It will come from [our grandkids] having grown up with games and emergent networks and who-knows-what-else that (obviously) we can't even imagine. They'll be able to navigate that stuff as effortlessly as we troglodytes can change radio stations while we're driving and talking to a passenger. In other words, we're not going to get much out of these vast data sets until we have people who can *intuit in six dimensions*. And then, watch out'[10] (emphasis mine). Sapolsky does not define what those dimensions are or what they look like, but I would like to explore what it would mean to 'intuit in six dimensions' and what it might mean for how we will interface with our electronic network of symbols.

At the moment, we gather in most of our symbols via our eyes. When we read a book, we activate the brain by the textual information we receive while we scan our eyes across the page. For all the wizardry the Internet represents, much of the information we receive through it is still in textual form, and we still need to scan our eyes over it in order to apprehend meaning (even if that reading path is associative and scattered). There is, of course, more visual information coming at us on the Internet as well, but consuming that information still requires us to scan our eyes over a scene to gather meaning, although this 'reading' process is very different from reading printed text. What if the symbols we receive from the Internet came to us via other senses, other entry points to and interfaces with the body, other ways of perceiving?

I imagine a scenario where we interact with such digital symbolic information via an expanded palette of our senses. There is reason to suspect that more information will come to us via our ears and from sound. By sound here, I mean much more than the human voice, although human speech will be an important way that we receive such information. Imagine having something like Google Glass in the ear? That is, what if the 'electronic voice in my head' that nudges me not to eat ice cream is actually in my head, speaking directly to my ear in the way that a cochlear implant allows the deaf to hear sounds? The ubiquity of ear buds today suggests that this scenario has already arrived.

Beyond the human voice, however, I believe it likely that we will see more data and information being 'sonified.' In the same way that we have visualized numerical and other kinds of information – that is, having made information that is not intuitively picture-like into pictures – we will hear more information as auditory sounds. To take our stock market example again, I can imagine a moment when such numerical information has been converted to sounds, the volume, pitch, and timbre of each tone a dimension of information about the state of the stock market. Such 'music' – such organized sound – would not be merely aesthetic, pleasing to the ear, but would hum in the background of our consciousness while we attend to other matters or turn our attention to other forms of information. Changes in this background music would signal changes in the market, shifting our attention and our auditory 'gaze.' But having information sonified in this fashion will allow us to add one more dimension of information to our cognitive palette, expanding our ability to perceive information beyond what only our eyes can take in.[11]

DOI: 10.1057/9781137460950.0007

I can imagine a day when symbols are 'olfactorized.' Via a similar process as data sonification, our symbols might be converted to smells that similarly hover in the background of our consciousness. Again, these smells would not be simply pretty odors, pleasing the senses as aromatherapy promises: such smells would represent symbols, data, and information, their intensity and fragrance corresponding to dimensions of information. Taking our stock market example, a pleasing scent like a flower might represent a booming stock market; the scent of rotting eggs would represent the bottom falling out of the market. The idea that smells might present us with meaningful data and information and that our symbols can be represented via scent probably strikes us as counterintuitive, even nonsensical. That would not be surprising, as 'smell is probably the most undervalued of the senses in modern Western cultures,' observes the designer John Thackara, undervalued in the wake of the Scientific Revolution and the Enlightenment.[12] Olfactorized information would once again elevate that sense as one capable of conveying symbolic meaning. 'Scent as symbol' would extend the capacity of the brain/mind/body to engage in cognition. Digitized scents are already being produced by the fragrance industries, and marketers already well understand the potential to brand products and experiences according to smells.[13] When information and data are similarly olfactorized, we will add another dimension to our growing sensory and perceptual system of symbolic manipulation: we will draw meaning and insight from all of our senses.

Other senses might be employed in an effort to aid our biological brains in gaining insight from information. What would it mean to 'feel information?' Thoughtful information designers have suggested that making digital information material through 3-D printing would allow us to touch and feel information. Information might also be transmitted to our skin through sensors that would alert us in the way sound or smell might. What would it mean to 'animate information,' to have movement as a dimension of information? Digital information already dances across our screens, but what if that movement were more deliberate, communicative, and meaningful, choreographed such that the movement of information itself was symbolically significant? The speed at which symbols move in front of us might represent a difference that makes a difference. As we have already seen when we pinch and swipe our screens, our movements already represent an important interface with the electronic symbolic storage system. We should anticipate that, like Tom Cruise's character in *Minority Report*, we will come to use our whole bodies to enact information and to interface with our symbols.

The limitation to our ability to make sense of the data we are producing is in part a function of the limited number of channels (or dimensions) of symbolic information we choose to pursue.[14] At present, the visual interface remains our chief means of receiving symbolic information. I imagine a scenario when we envelop/engulf ourselves in a perceptual sea of symbols, like the 3-D environment that Sapolsky mentions, extending our cognitive abilities across six dimensions, or more. 'Voice, haptics, and gesture will dominate the next era of interface,' predicts the Institute for the Future, 'but will soon be joined by responsive and anticipatory systems that will leverage implicit intention markers such as eye tracking, emotional signals, and other non-verbal cues.'[15]

In *The Watchman*, the villain Adrian Veidt observes world events via multiple television screens (this is the mid-1980s) and because of his superior intelligence is able to see larger patterns across these multiple screens, something like a 'dashboard' of information. Imagine a 'multisensory dashboard' that will send symbolic information to us: we will simultaneously read, see, touch, hear, even smell symbolic information, extending our ability to intuit meaning from the voluminous information that we are producing. The Tangible Media Group of the MIT Media Lab developed just such an environment, the ambientROOM, a designed space that translates information and symbols into sound, smell, temperature change, and other sensory interfaces that act upon the entire body.[16] The ambientROOM extends the memex beyond its screens, employing all of our senses as symbol-gathering interfaces beyond the books of St Jerome's study. The ambientROOM strikes me as just such a six-dimensional information environment anticipated by Sapolsky. The ambientROOM, like the study or the memex, could very likely become the interface with the Internet, the space that we enter in order to encounter the larger electronic symbolic storage system.

The ambientROOM – while sounding very science-fiction-like – resonates with earlier multisensory spaces. The first cave paintings at Lascaux and Altimira, for example, were not simply pictures drawn on walls. The caves themselves apparently were selected because of their acoustical properties, the manner in which sound was amplified or enhanced in these spaces, the sounds of galloping hoof beats to accompany the experience of the images.[17] In the scenario I am imagining, such digitized multisensory spaces would be filled with scents and sounds that would

DOI: 10.1057/9781137460950.0007

be as 'symbolic' as the words or images: that is, the cognitive interface (not simply the perceptual or emotional interface) would extend across the entire range of senses. The interface with the information flowing across the Internet would be a kind of digitized medieval cathedral: a multisensory interface of sound, image, and scent. Rather than saying that the environment will become more intelligent, we might instead anticipate that the spaces and points of contact with the electronic external symbolic storage system will expand to fill the environment around us. The cave, the library, the studio, the study, and the memex will not be spatially bound in a room but will expand across our surrounding environment.

At some stage, the multisensory space of the ambientROOM will be miniaturized and made ubiquitous such that it would take the form of a 'wearable computer,' an interface we would put on and take off as if it were a shirt, but would nevertheless connect us to the 'global brain' of the Internet. At that stage, we will carry the cave around with us, in our pockets, and we will change the logistics of knowledge.

If we believe that we can stave off information overload by digitizing information across all of our senses, then we will need to design systems such that our bodies will be able to coordinate all of the information coming through every sense. Initially, this will be overwhelming and bewildering: digital information will be arriving to us not just through screens or keyboards or a mouse but through a variety of interfaces operating upon our entire body. I wonder if such an avalanche of sensory information would be similar to the experiences of the first Europeans exposed to moving pictures. That is, we will be overwhelmed and disoriented at first, but we will eventually adapt to our new information surroundings through experience and education. To counteract what will no doubt be a sensory overload, we will need to learn how to interface with our multidimensional global brain. This might prove to be a new task of formal education: to teach us how to speak, to move, to touch, even to smell the information emanating from the electronic symbolic storage system.[18]

Indeed, interfacing with the Internet is redefining what it means to be educated. By this I mean that the Internet is altering our logistical relationship to knowledge. Merlin Donald argues that among the earliest civilizations, those that had developed writing systems, the sheer volume of external-to-the-brain symbolic materials they created necessitated a system of formal education.[19] One could claim that this has been the very definition of formal education ever since: that as humans accumulated more 'symbolically encoded things,' they required systems for acquiring, managing, manipulating, and demonstrating facility with those things. Education, in this formulation, is defined as the interface between our biological minds and the external storage system; education means learning to manage the dance between our biologically-based minds and the larger extension of our minds.

The philosopher of education Kieran Egan reminds us that 'one evident feature of our minds is that they are cultural organs. Humans have, for reasons that no doubt seemed evolutionarily good at the time, developed the means to store symbols outside our biological memory in such a way that we can access and retrieve their meaning at later times.'[20] This, of course, summarizes the extended mind hypothesis, but it is Egan's connection between these systems of external symbolic storage and its relationship to education that most interest me here. Again, Egan draws a connection between the external symbolic storage system and what we broadly term 'culture.' The relationship between the two is symbiotic: 'There is no mind in the brain until the brain interacts with the external symbolic store of culture.' If we may simplify: biological brain + external symbolic storage system (culture) = mind.[21]

Education, then, is the process by which the brain develops facility with that larger cultural system, the process by which we interface with the symbols stored outside the biological brain.[22] Cognitive tools in Egan's formulation 'are the things that enable our brains to do cultural work... These potentials of human brains are actualized only by the brain learning, and learning to use, particular pieces from our cultural storehouse. Culture, as it were, programs the brain.'[23] Egan is not talking specifically here about the Internet, but it should be clear from his definitions here that, as a system of external symbolic storage, the Internet will, of necessity, be folded into our systems of education, and will become one of our chief cognitive tools. How we maximize the Internet-as-cultural-storehouse strikes me as the central design issue for education in the 21st century. In what ways will we design formal education when cultural symbols are proximate to our location, when we need not be tethered to physical centers of the electronic symbolic storage system? How will we design formal education when our symbols are sonic and haptic, not just oral and textual?

The ubiquity of our interface with the Internet represents a change in our relationship to our cultural storehouse, a change in our relationship

DOI: 10.1057/9781137460950.0007

to knowledge. By 'relationship to knowledge', I mean a kind of logistic relationship between the individual human mind and the larger external symbolic storage system. For much of human history, the products of the external symbolic storage system were concentrated in relatively few locations. In the example cited earlier of the earliest civilizations, despite their growing volume, symbolically encoded things remained limited in location; to access the external system of symbolic storage, one needed to be in proximity to those great state libraries. Indeed, 'The Great Library' has long served as our principal metaphor for this extension of the human mind, the external symbolic storage system par excellence, our external 'cloud' that stores our symbolically encoded things outside our bodies. The interface with the Library has been one of the outstanding examples of how human beings have been able to extend our minds.

Among all of its effects, the emergence and maturation of the Internet has reconfigured the meaning of The Library, perhaps making this physical embodiment of the external symbolic storage system increasingly vestigial. This admission – which is difficult for me to state – carries with it a whole host of implications, especially about the changing relationship between our biological minds and this towering example of external symbolic memory system.

One could write the history of formal education as the spreading out of the external symbolic storage system from centralized locations. If we think of formal education as a logistical problem – how to locate externally preserved symbolically encoded things, how to access them, how best to manage them, how to demonstrate facility with them – then it is clear to see the potential implications of the Internet. In the first place, information and knowledge are starting to migrate from libraries, museums, and other physical repositories of knowledge outward into the 'cloud.' In doing so, our proximity to knowledge is changing, especially as our interface with that knowledge becomes more intimate. The Internet can be viewed as part of this larger narrative of spreading out knowledge and information from central locations, so that proximity to the sites of knowledge becomes less of an issue. Universal education is a historically recent phenomenon, and has been dependent upon, in part, the logistical question 'How do we make the products of the symbolic Cloud easily available to more people?' The printing press was one step in this direction: rather than being tethered to a large library or scriptorium at a monastery, one now had the ability to develop a personal library, a mini version of the external symbolic system. When Andrew

DOI: 10.1057/9781137460950.0007

Carnegie lavished millions to build public libraries across the country, he was similarly seeking to broaden the proximity to knowledge, to our system of external memory. The external symbolic system has always represented an external 'cloud' of symbolically encoded things outside our bodies, but that cloud has had a somewhat constricted shape. Throughout history, the configuration of the external symbolic system has altered, and there is reason to suspect that the Internet reflects the next stage of this larger reshaping of our external cultural systems.

To restate, the Internet reconfigures formal education by altering its logistics. That is, to be 'educated' still means to access, manipulate, manage, and to demonstrate facility with the external symbolic storage system. But how we will access and manipulate and manage and demonstrate facility will change. Of course, much of the information and knowledge migrating to the Internet is still read and viewed and experienced as it always has been. Accessing Shakespeare's plays via the Internet still requires one to read them; accessing a Picasso still requires one to look at the painting.[24] However, our proximity to these objects is now different, meaning that how we access these symbolic things has changed. As more and more symbolic objects move to the Cloud, as the logistics of access to information changes, the meaning of formal education is also changed.[25]

Academics have long situated The Library at the physical and conceptual center of the University. The current discussion about Google Books, for example, is as much about the future of the Library as much as anything. As the Library migrates to Google's Cloud environment, if The Great Library can be accessed from anywhere outside of Cambridge, Ann Arbor, and Oxford, what does this suggest about the logistics of the University? Do students still need to be proximate to cathedrals of learning? Proximity to knowledge will, of course, continue to matter. But it strikes me that the real revolution here is in the way formal education may no longer be required to be rooted in a specific place. This may explain in part why some academics worry about Google Books. Aside from copyright concerns – and I don't wish to overlook this very serious issue – I think some of the concern comes from the perceived conceptual hollowing out of the University. If the Library stands at its center and if that center is being dispersed into the Cloud, then will the University suffer the same fate as The Library? I understand that Universities are more than their libraries, and that much of the above are really symbolic statements, but the symbolism is important. The issues of proxemics and

DOI: 10.1057/9781137460950.0007

logistics associated with libraries can be equally applied to Universities. The current debates about online education – facilitated by the Internet – are usually fashioned as questions of access, that students can learn anytime/anywhere, and this is deeply troubling for some academics. We associate education as the meeting of a student and a professor, and have associated the physical proximity of teacher and learner as the most effective form of pedagogy. Those who reject online education do so, in part – because there other many other objections to be sure – because they believe that one cannot learn at an anonymous distance. (Of course, this assumes that books are poor pedagogical tools, because writer and reader are separated by both time and space.)

Advocates for online education assume that formal education need not be tethered to a specific location. While they do not use this specific language, these advocates are making the case that the diffusion of our external symbolic storage systems should continue to spread infinitely. Anya Kamenetz has been observing this movement of formal education into the Internet cloud. She draws upon her experiences with the TED Talks that are now easily accessible on the Internet, and observes that 'TED has become the new Harvard.'[26] Just as the symbolically encoded things contained in the Library are migrating to the Cloud, some of the best features of the University would also seem to have the potential to move to the Cloud. Our proximity to these great minds and great ideas is no longer tethered to place. The logistics of the University seem destined to be altered by the Internet.

We are not (yet) talking about an interface between the Internet and the brain's neural circuitry; the interface would be between our external symbolic storage system and our (bodily) systems of perception. Even if we start wearing Google Glass or have lens on our eyes, devices in our ears, sensors on our skin, the 'distance' between brain and external symbolic storage systems – however shrinking – would still be in place. The 'last mile' of that distance would be were we to directly connect – at the neural, synaptic level – the biological and electronic brains.

When we talk about the brain and the Internet being directly connected, I suppose we are haunted by the vision of the future represented in science fiction, from *Neuromancer* or the Matrix or the Borg. In each case, what terrifies us is the sense that the human part of that collaboration becomes vestigial: in the Matrix, the human body slumbers in stasis as the brain races through the virtual world, the shell of the human body is overwhelmed by the technological implants that gives the

DOI: 10.1057/9781137460950.0007

Borg their creepy pallid expressions. But, in one telling, our brains have always been 'connected' to the larger external symbolic system. If there is a more direct physical link, this might be unsettling, but it would be an advance on the metaphorical links that have long been established.[27]

Indeed, such direct human-machine interfaces already exist. Cochlear implants, for example, are small computers that wire directly to the brain of deaf patients, allowing them to hear. Early research on brain-computer interfaces demonstrates that, simply via a thought, some users can move a cursor around a screen or 'type' out letters. The initial applications for such interfaces are for quadriplegics; prosthetic devices are being developed that would allow those without the use of their limbs to nevertheless reestablish the links between brain and extremities that would allow them to have movement. For now, at least, we see such enhancements as useful for those with some physical impairment; we are not yet to the stage where a healthy adult will willingly allow such devices to be implanted, even if it were to enhance and amplify our cognitive abilities.

I compared the relationship between the brain and the Internet as potentially being like horse and rider, in that two intelligent systems would find a way to coordinate and choreograph their actions. There are moments when the horse must obey the rider and other times when rider must obey horse. As the brain and the Internet develop a more intimate relationship, the coordination of those two systems will represent one of the more significant design tasks of the near future. Michael Chorost argues that the human brain has already provided such a device to coordinate between two intelligent systems: the corpus callosum, that bundle of nerves that connects the two hemispheres of the biological brain. Chorost reminds us that the two hemispheres are separate cognitive systems; when surgeons are forced to sever the corpus callosum – necessary in some unfortunate cases to help patients with severe epilepsy – we discover that the two halves operate with separate functions and goals. It is as if our skulls actually contained two brains. The corpus callosum, the interface between the two brains, 'lets the hemispheres exchange so much data so quickly that functionally they behave as a unified brain.'[28] Chorost suggests that direct physical connections between brain and Internet would be the equivalent of an electronic corpus callosum: an interface between two cognitive systems. In effect, the electronic corpus callosum would connect our biological and technological hemispheres, these distinct 'brains' allowing for coordination between the two hemispheres similar to what occurs in our heads.

Both hemispheres would bring to the partnership their own strengths.[29] What will it mean to have two brains working together? Chorost sees a kind of cognitive division of labor, in the way our own brain is divided into hemispheres, that we will similarly have biological and electronic hemispheres, each assigned different tasks, yet working together (in harmony) to carry out cognition. Deep Blue is outstanding at chess, but cannot dance or fashion a painting. While the Internet may be fast, the brain is encased in a body, and as we indicated before, an important part of our cognitive abilities is that we move and interact in space. In such a partnership, our biological brains would become the body for the Internet. IBM's Watson can recall answers quickly and correctly, but cannot reflect on the meaning of its victory. The biological brain would be the reflective, introspective half of the partnership. Far from being overwhelmed or made vestigial by technology, our two brains – biological and technological, intimately interfaced – would work in partnership.[30]

Notes

1   Donald A. Norman, *The Design of Future Things* (New York: Basic Books, 2007), 9–10.
2   Norman, *The Design of Future Things*, 22. In 1960, computer scientist Man-Computer Symbiosis J.C. R. Licklider used the term 'symbiosis' to describe the relationship between humans and 'artificial intelligence.' J.C.R. Licklider, 'Man-Computer Symbiosis' *IRE Transactions on Human Factors in Electronics*, volume HFE-1, March 1960, 4–11.
3   Michael Nielsen, *Reinventing Discovery: The New Era of Networked Science* (Princeton: Princeton University Press, 2012), 113.
4   Nielsen, *Reinventing Discovery*, 114.
5   Eran May-raz and Daniel Laz, *Sight*, http://vimeo.com/46304267
6   On the design implications of information expanding across our environment, see Malcolm McCullough, *Ambient Commons: Attention in the Age of Embodied Information* (Cambridge, MA: MIT Press, 2013).
7   Andy Clark suggests that even such physical distances are not always present with interfaces, and that such intimate coupling has precedent. 'We discern an interface where we discern a kind of regimented, often deliberately designed, point of contact between two or more independently tunable or replaceable parts. It does not seem correct, however, to insist that flow across the interface be simple. The idea here seems to be that we find genuine interfaces only where we find energetic or informational

DOI: 10.1057/9781137460950.0007

bottlenecks, as if an interface must be a narrow channel yielding what [John] Haugeland describes as "low bandwidth" coupling. This is important for Haugland's argumentative purpose because he means to show that human sensing typically yields very task-variable, high-bandwidth forms of agent-environment coupling and thus to argue that no genuine interface or interfaces separate agent and world. Instead ... there is said to be "intimate intermingling of mind, body and world." But although agreeing with Haugland that sensing is at least sometimes best understood in terms of direct agent-environment couplings ... his own conclusion that no genuine interfaces then link agent and world seems premature ... An interface, I conclude, is indeed a point of contact between two items across which the types of performance-relevant interaction are reliable and well defined. But there is no requirement that such interfaces be narrow-bandwidth bottlenecks.' See Andy Clark, *Supersizing the Mind: Embodiment, Action, and Cognitive Extension* (New York: Oxford University Press, 2008), 32–33.

8   Robert Sapolsky, 'People who can intuit in six dimensions,' in John Brockman, ed. *This Will Change Everything: Ideas That Will Shape the Future* (New York: Harper Perennial, 2010), 368.

9   Martin Wattenburg, 'Map of the Market,' (1998) http://www.bewitched.com/marketmap.html

10 Sapolsky, 'People who can intuit in six dimensions,' 368–369.

11 See The Georgia Tech Sonification Lab. 'The Sonification Lab focuses on the development and evaluation of auditory and multimodal interfaces, and the cognitive, psychophysical and practical aspects of auditory displays, paying particular attention to sonification. Special consideration is paid to Human Factors in the display of information in "complex task environments," such as the human-computer interfaces in cockpits, nuclear powerplants, in-vehicle infotainment displays, and in the space program.' http://sonify.psych.gatech.edu/index.html

12 John Thackera, *In the Bubble: Designing in a Complex World* (Cambridge, MA: MIT Press, 2005), 178. 'According to the cultural historian Kate Fox, this was not always so: The current low status of smell in the West is a result of the "revaluation of the senses" by philosophers and scientists of the eighteenth and nineteenth centuries. The intellectual elite of this period decreed sight to be the all-important, up-market, superior sense, the sense of reason and civilization, writes Fox, "while the sense of smell was deemed to be of a considerably lower order – a primitive, brutish ability associated with savagery and even madness." The emotional potency of smell was felt to threaten the impersonal, rational detachment of modern scientific thinking.' See also Kate Fox, *The Smell Report: An Overview of Facts and Findings,* Social Issues Research Centre, http://www.sirc.org/publik/smell.pdf

DOI: 10.1057/9781137460950.0007

13   See for example Martin Lindstrom, *Brand Sense: Build Powerful brands Through Touch, Taste, Smell, Sight and Sound* (New York: Free Press, 2005); and C. Russell Brumfield, *Whiff! The Revolution of Scent Communication in the Information Age* (New York: Quimby Press, 2008).

14   Thackera, *In the Bubble*, 170–171. 'Sensitivity to changes in our environment through time develop best,' notes Thackara, 'if we learn to use all our senses, not just sight ... In reaction to the limited bandwidth of technology-enhanced vision, ecological thinkers emphasize that our senses ... are the fundamental avenues of connection between the self and the world,' although I would add that our senses will be similarly technologically-enhanced and that our environment will be as digital as it is physical.

15   'Internet Human | Human Internet Map,' *Institute for the Future* (May 2013), http://www.iftf.org/uploads/media/IFTF_TH12-InternetHuman_map_rdr.pdf

16   Craig Wisneski, Hiroshi Ishii, Andrew Dahley, Matt Gorbet, Scott Brave, Brygg Ullmer, and Paul Yarin, 'Ambient Displays: Turning Architectural Space into an Interface between People and Digital Information,' *Proceedings of the First International Workshop on Cooperative Buildings* (CoBuild '98), February 25–26, 1998, 23. The ambientROOM 'takes a broader view of display than the conventional GUI, making use of the entire physical environment as an interface to digital information. Instead of various information sources competing against each other for a relatively small amount of real estate on the screen, information is moved off the screen into the physical environment, manifesting itself as subtle changes in form, movement, sound, color, smell, temperature, or light.'

17   Thomas Rickert, *Ambient Rhetoric: The Attunements of Rhetorical Being* (Pittsburgh: University of Pittsburgh Press, 2013), 137–138. 'Prehistoric cave art combined sonics, graphics, and cave layout to define a place uniquely suited to a people's customs and rituals, an immersive "machine" to generate a sense of the uncanny, perhaps. Thus, sound and vision help constitute an ambient environment that induces and persuades through various forms of information flow and mood alteration. Like the use of images, the deployment of sound as afforded within the cave must be understood as a form of nonsymbolic communicative technology.'

18   I cannot help but to hear in those musings about a Global Brain doing our thinking for us the debates that mathematics teachers had (continue to have?) about the place of the calculator in mathematics education. When the first inexpensive handheld calculators began to proliferate, some educators argued that they were harming students' ability to acquire mathematical knowledge. Since school kids could simply punch in numbers, they no longer needed to memorize multiplication tables. Other educators argued that calculators freed students from such rote drudgery, meaning that

DOI: 10.1057/9781137460950.0007

teachers could move on to advanced mathematics concepts, ultimately improving the quality of mathematics education. Does a calculator do our thinking for us? Or is it a tool that enhances our thinking, a technology of external memory storage that can only be activated by our biological brain?

I suspect we will have the same sorts of discussions about the just-in-time external knowledge storage system represented by the Internet and the effects on education. Evoking the Doonesbury character who does not hold knowledge in his biological memory but who resorts to Google for answers to the professor's question, there will no doubt be teachers who lament how little our students are remembering and how dependant they have become on the Internet for their memories. Indeed, in such a scenario, it is easy to understand why some would wonder if the Internet were doing our thinking for us. But I suspect there will be a growing number of teachers who will fashion a new pedagogy focused not on rote memorization but on the new kinds of thinking that can be 'freed up' once rote memorization is replaced by this globe-spanning technology of external memory storage.

19 Merlin Donald, *Origins of the Modern Mind: Three Stages in the Evolution of Culture and Cognition* (Cambridge: Harvard University Press, 1991), 320. 'Compared with the monotony and redundancy of the hunting-gathering lifestyle,' he writes, 'these early centers of graphic invention exploded with symbolically encoded things to be mastered. Large state libraries were already a reality in ancient Babylon, and by the time of the Greeks [external symbolic storage system] products had been systematically collected and stored in several world centers of learning. At this point in human history, standardized formal education of children was needed for the first time, primarily to master the increasing load on visual-symbolic memory. In fact, formal *education was invented mostly to facilitate use of the ESS*.'

20 Kieran Egan, *The Future of Education: Reimagining our Schools from the Ground Up* (New Haven: Yale University Press, 2008), 38.

21 Egan, *The Future of Education*, 41. 'The mind is nearly invisible in our current understanding of the brain,' contends Egan, 'but it is very plainly visible when viewed in terms of the culture our brains interact with.'

22 Egan, *The Future of Education*, 40. 'If we see this external symbolic storehouse as something whose internalization in individual brains constructs their minds, and if we accept Vygotsky's idea that the tools, or "operating systems" and "programs" for our brains initially exist external to our bodies in our culture, then we may begin to conceive of education's tasks somewhat differently. Education becomes the process in which we maximize the tool kit we individually take from the external storehouse of culture. Cultural tools thus become cognitive tools for each of us.'

23 Egan, *The Future of Education*, 41.

DOI: 10.1057/9781137460950.0007

24   I understand, of course, that there are also new ways to 'read' and 'look' facilitated by the affordances of digital technology.

25   I am reminded of my experiences teaching history at Heidelberg College in the small town of Tiffin, Ohio. I was asked to teach the historical methods course, which required a substantial research paper from each student based on archival research. As you might imagine, there are relatively few archives in and near Tiffin, thus limiting the kinds of research my students could undertake. When I made online archives part of the assignment, like the collections of the Library of Congress, my students could now expand their potential research subjects. This new proximity to information is already having profound effects on the way we think about history education at the K-12 level. There is a growing push for students to actually 'do history' in middle and high school history classes, meaning engaging in primary source research of the kind all historians engage in, primary source research being a more authentic way to learn history. This vision of authentic history teaching depends upon proximity to primary documents, usually physically stored in archives and large libraries, a logistical problem for teachers serving in small towns and other areas at some distance from repositories. Making archival materials available on the Internet reduces the proximity issues. This one narrow example demonstrates how the logistics of knowledge are altered by the Internet, with enticing potential effects for formal education.

26   Anya Kamenetz, 'How TED Became the New Harvard,' *Fast Company* 82, September 2010, http://www.fastcompany.com/1677383/how-ted-connects-idea-hungry-elite. 'If you were starting a top university today, what would it look like? You would start by gathering the very best minds from around the world, from every discipline. Since we're living in an age of abundant, not scarce, information, you'd curate the lectures carefully, with a focus on the new and original, rather than offer a course on every possible topic. You'd create a sustainable economic model by focusing on technological rather than physical infrastructure, and by getting people of means to pay for a specialized experience. You'd also construct a robust network *so people could access resources whenever and from wherever they like*, and you'd give them the tools to collaborate beyond the lecture hall. Why not fulfill the university's millennium-old mission by sharing ideas as freely and as widely as possible?' [emphasis mine]

27   Andy Clark would contend that we have long been cyborgs: 'We some of the "cognitive fossil trail" of the cyborg trait in the historical procession of potent cognitive technologies that begins with speech and counting, morphs first into written text and numerals, then inot early printing (without moveable typefaces), on to the revolutions of moveable typefaces and the printing press, and most recently to the digital encodings that bring text, sound, and image into a uniform and widely transmissible format. Such

DOI: 10.1057/9781137460950.0007

technologies, once up and running in the various appliances and institutions that surround us, do far more than merely allow for the external storage and transmission of ideas. They constitute, I want to say, a cascade of "mindware upgrades": cognitive upheavals in which the effective architecture of the human mind is altered and transformed.' Andy Clark, *Natural-Born Cyborgs: Minds, Technologies, and the Future of Human Intelligence* (New York: Oxford University Press, 2003), 4.

28   Michael Chorost, *World Wide Mind: The Coming Integration of Humanity, Machines and the Internet* (New York: Free Press, 2011), 9.

29   Chorost, *World Wide Mind*, 8.

30   Daniel Pink has written about the increasing importance of 'right brain' skills, which he calls 'R-Directed aptitudes.' These include design, synthesis, play, and meaning-making. As the internet hums along powered by algorithms operating on data, it could very well develop into a 'left brain' that engages in analysis, while the biological 'right brain' engages in R-Directed aptitudes. We would, in effect, off-load the left-brain to the algorithms of the Internet. See *A Whole New Mind: Moving from the Information Age to the Conceptual Age* (New York: Riverhead Books, 2005). Also note that Daniel Kahneman, , and his notion of System 1 (unconscious processes) and System 2 (conscious processes). Both systems co-exist inside the skull, which is suggestive of how the Internet 'system' and the biologically-based brain 'system' might cooperate and interact. See *Thinking, Fast and Slow* (New York: Farrar, Straus and Giroux, 2011).

5

Limit

Abstract: *That the Internet will become an autonomous partner in cognition is far from inevitable. What appears as an inexorable process can be driven off-course when it collides with countervailing trends. Will we reach the physical limits of computing power, and is the growth of the Internet itself subject to physical limitations? Can we be assured that the infrastructure of the Internet will be maintained? Does the brain have a physical carrying capacity beyond which it is unable to engage in coupled cognition with the Internet? Will we encounter the limits of our ambitions, and abandon the dream of autonomous cognition and brain–Internet interfaces? What will result when our impulse to extend cognition via the Internet confronts any number of limits upon that impulse?*

Keywords: future; information; Internet; networks; physical limits

Staley, David J. *Brain, Mind and Internet: A Deep History and Future.* Basingstoke: Palgrave Macmillan, 2014. DOI: 10.1057/9781137460950.0008.

DOI: 10.1057/9781137460950.0008

Futurists in the 1970s might have been excused for their giddy predictions about the future of space travel, of the real possibility of the colonization of the moon and the expansion of the human species across the solar system. After all, the Apollo missions had successfully landed men on the moon such that those moon landings were seemingly commonplace. One only had to draw the trend lines forward: Apollo's success, American technological and engineering capabilities, and a limitless desire to expand meant that colonies on the moon were sure to happen by the end of the millennium.

Of course, we have yet to colonize the moon. There have been a number of factors that have limited this 'inevitable' expansion: waning interest from politicians and the general public, problems at home that demanded our attention and resources, a post-Vietnam/post-Watergate malaise about the American spirit, a thaw in the Cold War that dampened the driving force behind the moon landings. That is, the trends pointing toward moon colonization ran up against blockers or countervailing trends that pushed that trend line in an opposite direction from the predicted path. (Indeed, with the retirement of the last space shuttle, the future of the US-manned space program is in limbo today. It is possible that the colonization of the moon will be carried out by private companies, not government-funded space agencies.) Those futurists in the 1970s did not fully apprehend the limits that could constrain the advance of trends.

Thoughtful futurists today understand that the future cannot be predicted with certainty. Because of sensitivity to initial conditions, we cannot know the precise state of a system at any given future n-point. Variables interact in complex ways and trends can negate advances in one area such that there cannot be one certain path to the future. Today's futurists do not think in terms of predictions, but rather in terms of multiple scenarios: a number of narratives that each describes different, equally possible, future n-states.[1] Futurists will often explore multiple scenarios to better understand the ways in which trends and drivers might interact, and to better understand the limits under which trends might unfold.

In the previous two chapters, I developed scenarios of the future of the Internet that were based on a number of assumptions: that computing power will continue to expand exponentially such that our digital network would begin to mimic cognition, that the infrastructure of the Internet will remain robust and well-maintained to enable such cognition,

DOI: 10.1057/9781137460950.0008

that we will remain comfortable with storing our symbols externally in evanescent digital form. That the Internet will become an autonomous partner in cognition with the human brain is far from inevitable; like the colonization of the moon, what appears as an inevitable and inexorable process can be derailed or driven off-course when these trends collide with countervailing trends.

Ray Kurzweil's vision of an artificial brain that will seamlessly integrate with the biological brain is based on the continued logarithmic expansion of Moore's law which, it must be remembered, is not so much a law as it is an assumption about how processing power and speed will continue to double. Some physicists have observed that we will eventually reach the physical limits of computing power, and that Moore's law won't operate indefinitely since we will reach the limits imposed by the material objects needed to conduct computations.

The physicist Michio Kaku notes that we currently etch transistors onto silicon chips, and that because we have been able to narrow the beam of light necessary to conduct the etching we have been able to make chips smaller and smaller. But this process cannot continue indefinitely. At some point very soon we will reach a state where the laws of physics will operate, and we will not be able to make etchings smaller than atoms. We will have run up against the limits of Moore's law, which Kaku estimates will occur around the year 2020.[2] Even Gordon Moore conceded that constantly doubling computing power could not carry on indefinitely.[3]

As we approach smaller and smaller sizes for transistors, eventually the realities of quantum mechanics will dictate terms. Because we cannot know the position or speed of electrons at the same moment, we will reach a stage when we will not be able to contain the electrons, Kaku noting that these will 'leak out' of any device that we might try to construct. This is not to suggest that physicists and engineers are not attempting to solve this problem, and have posited solutions such as quantum computing, atomic transistors, and DNA computing. Each has its promises, but also has both physical and economic limitations that make their widespread development and application far from inevitable.[4] Of course, humanity has often found ways to break through seemingly impenetrable boundaries, and the physical limits of computing may be one of these. However, these limitations warn us that we cannot assume ever-greater computing power and the artificial cognition it might engender.

DOI: 10.1057/9781137460950.0008

The growth of the Internet – necessary for the kinds of autonomous cognition envisioned in the previous scenarios – might also be subject to physical limitations. Jeff Stibel contends that all networks are bound to what he terms the 'network curve.' The curve is a three-stage process: a period of rapid growth that extends the size of the network beyond a physical carrying capacity, at which point the growth curve levels at a stage that he calls the 'breakpoint.' The size of the network then retreats across the threshold of its carrying capacity and settles back into an equilibrium. Stibel, the brain scientist, maintains that this network curve is a law governing all networks. The brain, being a network, also grows, hits a breakpoint and settles into equilibrium. Since the Internet is also a brain-like network, it too is heading for a breakpoint, reaching a physical limit beyond which it will no longer grow and retreat into a stable equilibrium. Networks of all kinds develop a kind of intelligence once they hit the equilibrium stage, and Stibel believes that the Internet, once it reaches its mature equilibrium phase, will similarly develop a kind of intelligence.[5]

It may be the case that the Internet will level off into an intelligent system. But Stibel has introduced here the possibility that the Internet cannot grow indefinitely. It is also possible that the intelligences and cognitive abilities we may ascribe to networks may not be as robust as might be assumed from the other scenarios. That is, a leveling-off of the growth of the Internet might just as likely represent a leveling-off of any cognitive behaviors we might assume it will acquire. What will happen as we approach the limits of the size of the Internet? Like other ecological systems, are there 'Limits to Growth' to the Internet? An ever-growing Internet might simply not be sustainable. The Internet might reach limits, and thus remain nothing more than a very powerful semi-autonomous exogram.

An Internet of Things scenario and its concomitant 'secret life of data' scenario both assume a robust and well-maintained Internet infrastructure. I once observed that we sometimes think about the Internet as if it were a Gothic Cathedral: we marvel at the beautiful stained glass and allow the light and acoustics to wash over us, while ignoring the ugly flying buttresses that hold the structure in place. Our scenarios about the future of the Internet assume the continued maintenance and stability of those electronic flying buttresses, and this is an assumption that is far from certain. We know that at this moment our own physical infrastructures of roads, bridges, and airports are crumbling and that we apparently

DOI: 10.1057/9781137460950.0008

lack either the resources or the political will to repair them. Will the same be said about the physical infrastructure of the Internet? Even if the infrastructure of the Internet is well-maintained, critics alarmed at the prospect of the end of 'net neutrality' anticipate a scenario where Internet cost and service is uneven, with some users enjoying better benefits than others. Worse, without net neutrality, information may not flow as freely across the Internet as we have come to expect.[6] Autocratic governments have frequently shut down or disrupted the Internet as a way to control the flow of information. What might such disruptions look like in the future, and how might they impact any coupled cognitive activities dependent upon the Internet? The scenarios developed in Chapters 3 and 4 both assume a robust and stable network structure across which symbols and information flow freely and uninhibited. Can we be assured that the infrastructure of the Internet will be maintained such that it will allow for the futures we have assumed?

Some futurists have warned of an impending energy shortage in electricity.[7] Because of an ever increasing demand for electricity – from electronic gadgets and, potentially, electric-powered cars – Western countries could see the kinds of daily power outages and rationed of electricity that we associate with postwar Baghdad. Cyberwar and cyberterrorism are quickly becoming daily realities, with rogue elements hacking into and disrupting electronic networks. How does electronic cognitive off-loading work when electricity and the reliability of electrical grids are so potentially unpredictable?

Other futurists have posited that sunspot and solar flare activities could disrupt or even wipe out the electronic grid, and with it the digital memory of our external symbolic storage system. In 1859, an immense solar flare – subsequently called the 'Carrington Event' after the astronomer who witnessed it – engulfed the earth such that auroras were viewed even in tropical latitudes. The Carrington Event also wreaked havoc on the nascent telegraph system. As long as that system was in its early days, and before electronic communications became so globe-spanning, its impact was relatively minor. Today, of course, the electronic network is much larger and wider, much more developed, much more intrinsic to our thought and communication. A Carrington-like event today would be devastating.[8] It does not take an active imagination to foresee the possibility of another Carrington-like event, one that would disrupt and disable the Internet such that it would be inoperable. How will cognition be disrupted when the electronic half of our distributed cognitive

DOI: 10.1057/9781137460950.0008

system is so unstable? What happens when our digital external memory can no longer function as a reliable extension of our cognition? Clearly, an external memory system like the Internet dependent upon electricity has many – at this stage only remotely possible – vulnerabilities, with an electronic 'burning of the Library at Alexandria' a potential threat.

Even if the Internet is not disrupted by corporate greed, political manipulation, calamity, or physical limits, what does 'memory' look like when digital information appears fleeting and evanescent, with a dramatically reduced shelf-life? If so much information and knowledge, so much of our memory and cultural symbols, is ascending to the electronic cloud, one potentially ominous concern – more so than the question of whether or not the Internet is making us stupid – is the seemingly volatile shelf-life of this memory. As an historian, I come from an academic tribe that values The Book in part because, as a storage medium, books seem to be relatively permanent objects. If I publish a book today, there is every reason to anticipate that the physical object will be around hundreds of years from now. For scholars who value the past, this information longevity is an important facet of knowledge and knowledge creation.

But we need not focus on doomsday scenarios to understand the volatility of electronically based information. I like to tell my students the story of my dissertation: while you can read a printed copy of my work, completed in 1993, the electronic version is harder to access. My 250-page dissertation is stored on seven (yes, seven) 5.25 inch floppy disks, and so one would need to have access to a computer with a 5.25-inch drive to read them. The dissertation was written in WordStar, which was an obsolete program even in the early 1990s, but must be even more difficult to find today. Even after 20 years, this electronic information is very difficult to access.

This remains one of the more significant challenges of the Internet for librarians, archivists, and other curators of digital information. Once information has been electronically encoded and uploaded to the Internet, who is going to maintain it and for how long will it be maintained? Who will pay for this maintenance? Perhaps more importantly, why should such information be maintained? The Library of Congress has recently agreed to begin archiving Twitter feeds, but as the curators are well aware archiving in an electronic environment is not the same as archiving books. As new versions and upgrades of familiar applications inevitably come along, how easy will it be to migrate electronic

DOI: 10.1057/9781137460950.0008

information to these new environments? There is reason to be cautious about information in the electronic cloud being as difficult to access as my 1993 dissertation. It is possible that, despite the assurances of librarians and archivists, the vast amount of digital information cannot be permanently stored and preserved. We may discover that it is simply too costly, too time- and resource-intensive, to upgrade and maintain electronic data. When we reach such limits to our capacity to preserve electronic data, our information will eventually be lost, and this moment could be arriving sooner rather than later. For there to be a secret life of data and all of the autonomous cognition that 'algorithms acting upon data' promises, there also needs to be a reliable way to maintain that store of data. It is possible that there are limits to what we can realistically be expected to preserve.

The media theorist Wolfgang Ernst in fact argues that 'the Internet is not an archive.' Because information transmitted across the Internet is unstable because it is rewritable, this instability makes the notion of the Internet as archive untenable, since an archive implies permanent storage of stable cultural artifacts. If information on the Internet is unstable, this would suggest that we may come to question its value as an external symbolic storage system.[9] What if more and more of the information that is created for the Internet is deliberately understood to be fleeting and short-lived, like a mayfly? That we begin to think of the information and symbols that flow across the Internet not in the language of permanent storage or archival selection, but in the language of 'end-of-life' issues? What if we decided not to preserve our digital works? Museums save and preserve physical objects, but do not save or preserve exhibitions. Is the Internet a kind of giant electronic exhibition? Can we plan now for a time in the future when scholarship, information, and knowledge are understood as ephemeral performances? What are the protocols of acceptable loss of digital information? That we were even having this conversation suggests that in the realm of the Internet there is less of a concern with long-term preservation. The Internet may foster a here-today, gone-tomorrow approach to information and knowledge, information as fleeting and temporary, more temporally situated than the book and the library.

Creating information on the Internet becomes the gesture of a historical moment; when we say that knowledge and information on the Internet is 'just-in-time,' we might also refer to its ephemeral, at-this-moment quality. What kind of meaningful symbolic storage does the Internet

DOI: 10.1057/9781137460950.0008

represent? I think a more important question to ask about the Internet is not is it making us stupid, but is it making it possible for us to forget?

A potential and countervailing trend might be a larger societal 'exhaustion with the evanescent.' We may grow tired of the volatility of information and long once again for physical, tangible artifacts as the basis of our external symbolic storage system. Such a scenario might seem fantastical, given the assumptions and beliefs of our current digital moment, but there are some faint signs of a 'return to the material,' to borrow from the name of a recent conference.[10] Indeed, the growing popularity of 3-D printing might signal a cultural re-turn toward making physical objects rather than consuming digital bits. Imagine a scenario where symbolic objects begin in material form, are converted to digital signals that sit in electronic stasis until they are 'rematerialized' by a user elsewhere in physical form once again. (Think of the way that many users still print off copies of electronic documents rather than allowing these to remain in evanescent digital form. Imagine the same impulse only with physical objects, not just sheets of paper.) Should 3-D printers advance in speed and their ability to replicate matter – a big assumption – it is possible that users will express a preference for information and cultural symbols in physical, tactile form. The digital, then, would be only a transit, not the final form of information, a way to transport symbolic objects. Such a scenario emerges, however, only if our culture reaches the limits of its tolerance for evanescent digital information.

Assuming that our culture of print remains a vestigial part of our cognitive architecture, and in the same way oral cultures sometimes re-emerge after a period of literacy, we might be surprised to discover a robust return to a print culture that never really disappeared. (This would be similar to our current cultural moment, where vinyl records – assumed to have been made obsolete by CDs and MP3s – have made a comeback.) For all of the predictions about the end of the book, physical books remain viable, even vibrant, forms of external symbolic storage, and there are reasons to suppose that they will remain so into the future.

We have been considering the potential physical limitations to the technical hardware of our coupled cognitive system, the limits to both computers and to the networks that link those computers together. The other part of this coupled system – the biological brain – might also exhibit physical limitations that would render the 'Query' and 'Interface' scenarios far from inevitable. As the information and symbols that pass

DOI: 10.1057/9781137460950.0008

through our digital networks grow, we could potentially encounter limits in our biological brain's capacity to make sense of that information. Indeed, one study suggests that we are already running up against such physical limits. Using the Weber-Fechner law, which posits a logarithmic relationship between stimulus and perception, researchers at the Institute for Theoretical Physics at Goethe University maintain that a similar law governs our capacity to absorb more information. 'The neuropsychological capacity of the human brain to process and record information may constitute the dominant *limiting factor* for the overall growth of globally stored information,' conclude the researchers (emphasis mine). The suggestion is that, given that our digital networks contain ever-growing amounts of information, there are theoretical limits to the meaning that we will be able to apprehend from that information.[11]

In the novel *The Circle*, David Eggers describes a near-term future where brains are pushed beyond what would seem to be a realistic carrying capacity. Employed as a 'customer experience' representative to the mega-corporation called The Circle, the protagonist is introduced to her work station. She is faced with three screens, one that takes in customer queries, one that displays messages from her supervisors, and a third screen containing feeds from both an internal and an external social network. All have to be monitored and responded to as part of her work detail. She eventually adds another device that 'speaks' in her ear, allowing her to respond to even more queries and to update ever more social networks with a simple voice response.

The protagonist becomes adept at managing all of these digital channels such that she quickly rises within the company. But the description of her interface with the digital network seems fantastical (even for a futuristic novel). Consider this scene:

> Mae looked at the time. It was six o'clock. She had plenty of hours to improve, then and there, so she embarked on a flurry of activity, sending four zings and thirty-two comments and eighty-eight smiles. In an hour, her PartiRank rose to 7,288. Breaking 7,000 was more difficult, but by eight o'clock, after joining and posting in eleven discussion groups, sending another twelve zings, one of them rated in the top 5,000 globally for that hour, and signing up for sixty-seven more feeds, she'd done it. She was at 6,872 and turned to her InnerCircle social feed. She was a few hundred posts behind, and she made her way through, replying to seventy or so messages, RSVPing to eleven events on campus, signing nine petitions, and providing comments and constructive criticism on four products currently in beta.[12]

Later on in the novel, Mae joins the CircleSurvey and averages 1,345 questions a day, this on top of all of the other feeds described above and others not listed here. Obviously, this is a work of fiction, but I am nevertheless drawn to Eggers' imaginative projection because it forces us to consider the physical limits of our brain's capacity to meaningfully interact with a growing electronic external symbolic storage system. Mae frequently attends to multiple digital feeds, although we are not really shown how she does this, only being told that she 'makes her way through.' The reader is left with the suspicion that no one person could possibly manage that much information; the descriptions of Mae's encounters with zings and smiles and feeds and the Inner- and OuterCircle become more and more implausible. Are there enough minutes in an hour and hours in a day to attend to that volume of information? The purpose behind Eggers' claustrophobic description of a brain overwhelmed by digital information is to ask the reader 'Is this any way to live?' (Eggers contrasts Mae's working life with her quieter and clearly more desirable moments spent alone kayaking, or with her parents' simpler, technologically unmediated and more joyful life.) That question does occur, but, for purposes of this essay, an even more pertinent question might be 'Is this life even theoretically possible?' For narrative purposes, Mae, as with the other high-achievers at The Circle, manages continual and insistent amounts of digital information. But I wonder if Mae's cognitive management is a narrative conceit, like traveling faster than the speed of light, a way to skirt around the laws of physics in order to make a larger point.

Does the information and symbolic manipulation ability of the brain/body have a carrying capacity? Think of runners of the 100-meter dash. Over the past century, times for the Olympic champion have been getting shorter and shorter. But those times, when plotted on a curve, seem to be approaching some sort of asymptotic limit, leaving some to wonder if there lies a physical boundary, no matter what other kinds of enhancements we might apply, beyond which we will not be able to move our bodies any faster.[13] Does the brain, even one enhanced by the Internet, similarly have a physical boundary beyond which it is unable to meaningfully engage in coupled cognition with the electronic network?

What would result if we did reach such a theoretical carrying capacity? Would the electronic symbolic storage system similarly stop growing? Will the limits of the brain thus constrain the size, scale, and cognitive autonomy of the Internet? Stanislas Dehaene identified the limits the reading brain has established around the graphical marks in

DOI: 10.1057/9781137460950.0008

all writing systems. Will the limits of the brain similarly constrain the Internet? Symbolic technologies have no doubt expanded the capacities of the biological brain, and that brain establishes the possibilities for the development of further symbolic technologies. But that brain also establishes limits, and reins in the actualities. Will the brain similarly establish limits around the amount of information we produce, on the size of our electronic symbolic universe, on the shape of the Internet? If the conclusions reached by the researchers at Goethe University are borne out, it suggests that the future growth of the Internet and the electronic symbols that flow across it will be limited by what the biological brain can handle.

In Chapter 4, I noted that Robert Sapolsky said that to increase our information carrying capacity, we will need to be able to intuit data in six dimensions. I imagined a scenario where we would begin to receive digital information in the form of sounds, smells, and tactile sensations, meaning that in order to ingest ever-growing amounts of information that we would need to expand the sensory apparatus through which we acquire information, that the interface with our symbols would arrive through all of our senses. Even if we are able to manage the olfactorization of information, is there a point where our capacity to query symbols will be reached, and that no number of alternate channels will extend that capacity? Is there a limit to the amount of symbolic information the brain and body is able to ingest? Even with such an expanded palette, will we reach limits such that our bodies will not be able to adequately function as a partner in a coupled system of cognition?

If our brains and bodies reach such a theoretical boundary, will the Internet simply continue on without us? Consider the 'secret life of data' discussed in the previous chapter, the notion that cognition, in the form of pattern recognition on the part of algorithms, can carry on outside of our biological brains. Will we entrust to our algorithms some of the function of making sense or of drawing insights from data, information, and symbols? I note that the firm Narrative Science is already doing something like this.[14] Narrative Science engages in data-mining, in that their algorithms comb large amounts of data seeking meaningful patterns. Instead of producing a spreadsheet or table of data, Narrative Science produces written texts. Some have suggested that we would no longer need business reporters to write journalistic accounts, when our algorithms could produce those narratives for us.

DOI: 10.1057/9781137460950.0008

This scenario assumes that there is a human brain that is reading those narratives produced by the algorithms. Given the limitations the brain may impose, would we reach a stage where even texts produced by computers would approach a volume such that it would not be possible to read them all? (The world's libraries today groan with millions of volumes that one person could not be expected to read.) It is possible that once the physical limits of the brain have been reached, that computers and the networks that link them together will go on producing – and reading for themselves – those narratives without the intercession of humans. Peter Swirski imagines just such a scenario, although he is quick to point out that we are already living under such conditions. A good amount of the trading that occurs in the stock market, for example, is done by computers without any human input. Swirski evokes Stanislaw Lem's concept of 'bitic literature' (or 'biterature,') produced by 'computhors,' a situation where algorithms are producing semantically meaningful texts. (Again, we already appear to have reached that stage, if Narrative Science's business model is any indication.) A computer that produces a text such that, when read by a human reader, convinces that reader that another human wrote it would pass the Turing's test for intelligence. We may well ask whether a text from Narrative Science would pass such a test. Swirski raises another possibility: that eventually the works of computhors will be read only by other computers, algorithms reading the works of other algorithms. In such a context, a language community would emerge that would be devoid of any human participation. And why not: 'Human sensory and thus information-bearing capacities have, after all, remained essentially unchanged for at least the last 100,000 years... [and yet] the quantity of information online far outstrips what any non-artificial intelligence could fathom.'[15] In such a language community, a new kind of Turing's test might manifest: a computer convinced that what it was reading was produced by a human, when in fact it was produced by another computer. If we were to reach the stage where computers would produce information to be read by other computers, we will have removed humans entirely from the coupled cognitive system. A discussion among algorithms might occur without our even knowing it is happening. Or we would witness the symbolic activity, but have no way to interpret or make sense of what is happening.

This theoretical possibility represents the kinds of cognitive activities that might occur on the other side of the brain's carrying capacity. We

DOI: 10.1057/9781137460950.0008

would have reached a bifurcation point: if we were to arrive at a boundary beyond which the brain cannot read or interpret any more external symbolic information, then the Internet either stops growing, or the Internet will speed past that limit, traveling on without our participation. It is possible that the path the future follows at this bifurcation point will not be made deliberately by either designers or users, and that the limits of the brain will determine the limits of the Internet. There will be some neurological and physiological limits to the expansiveness of the Internet and its interface with the human brain that even designers, technologists, entrepreneurs, and others attempting to increase the size and scale of the Internet will not be able to transgress.

It is also possible that we will run up against the limits of our ambitions. Jaron Lanier, for example, is one such technologist who has had a change of heart. One of the pioneers of the wired world and 'father of virtual reality' now scolds the culture of 'computerism' and the Silicon Valley start-up community for bad design, an attitude that he says is anti-human, that worships the noosphere. He is especially troubled by the false 'wisdom of crowds' that the Internet has unleashed. 'Different media designs stimulate different potentials in human nature,' Lanier counsels. 'We shouldn't seek to make the pack mentality as efficient as possible. We should instead seek to inspire the phenomenon of individual intelligence.'[16] Lanier draws to our attention that the development of the Internet has been concentrated in the hands of a few influential designers, technologists, and entrepreneurs. The design choices made by this group have set all of us along a course that Lanier – one of those influential designers – now regrets. He observes that these designers have imposed their designs upon unsuspecting, or at least unconsulted, users. Going forward, at a minimum, designers of technological systems must work with users: intent must be embedded as much by users as it is by coders and software engineers.[17]

What if all of these influential designers and technologists were similarly converted to Lanier's way of thinking? Would they send the development of the Internet along a different path, one that 'inspires the phenomenon of individual intelligence?' Perhaps these designers will become apostates to the church of the Singularity, and turn away from the goal of building an autonomous brain and toward some other goal. If these designers and entrepreneurs are as influential in shaping the course of the future as Lanier claims, then their future behavior will influence the direction of the Internet. They may collectively choose not

DOI: 10.1057/9781137460950.0008

to continue to develop systems featuring implantable interfaces with the Internet as an autonomous intelligence.

If that scenario seems unlikely, consider the change in direction many designers took over the course of the 1970s. Designers in the 1950s were complicit in the system known as 'dynamic obsolescence,' meaning changing the look, style and appearance of products as a way to create new demand for the product. In the 1970s, designers such as Victor Papanak chided designers for these practices, calling them environmentally unsustainable. Generally speaking, designers today are embracing instead sustainable design, and would find 'dynamic obsolescence' unsustainable.[18] It is possible that tomorrow's designers and builders of the Internet will be influenced and inspired by technologists such as Lanier, and jettison the pursuit of an artificial neo-cortex or to redirect their efforts away from building a cognitively autonomous network.

I suspect that end users are not as impotent in this process as Lanier suggests, such that they need to be invited into this design process. End-users and the larger society play a critical role in technological development. Brian Winston has demonstrated that technologies may indeed be developed by technologists and designers, but unless these are deemed useful and adapted by the wider society – a process he terms 'supervening social necessity' – they remain merely interesting prototypes, confined to the labs in which they were developed.[19] Lanier, Carr, and other critics see the trends pushing us toward wearable computers or artificial cognition coming from misguided engineers, corporations, or other nefarious groups. It is possible that the future of the Internet will be determined by the actions and behaviors of users, and that these users could decide to unplug, to interface in other ways, or to de-interface entirely. Google Glass might remain an interesting prototype, for 'society' may determine that no practical reason exists to adopt this new tool.

The secret life of data scenario assumes that users will continue to exhale volumes of information that will be captured and read by algorithms. The rise of whistleblowers such as Edward Snowden, however, could signal a trend toward the scaling back of big data, and especially of users pushing back against the expansion of such data mining efforts. Users might collectively decide that the 'benign surveillance' that Big Data represents is too repressive, and would clamor for measures that would severely curtail Big Data collection, As I write this, the European Court of Justice has ruled that Google, when requested, must remove links to and otherwise stop the search for 'irrelevant' user information.

DOI: 10.1057/9781137460950.0008

In this particular case, a man who was in financial difficulty put his property up for auction in 1998. While his fortunes have since improved, a Google search of his name nevertheless continues to announce his past economic difficulties. Advocates see this case as a victory for what is being termed 'the right to be forgotten.'[20] At a minimum, such a court ruling suggests that users have rights over and can legally control the use of their own data. One of the main assumptions of the secret life of data scenario is that individuals secrete data (perhaps without their knowledge), data that is captured and analyzed for patterns that reveal behavioral traits and other patterns such that decisions and 'nudges' can be designed. If more and more individuals exercise their right to be forgotten (or to be digitally anonymous), how efficiently will an autonomous cognition based on 'algorithms acting on data' be able to function? As the EU case demonstrates, there may very well be a 'data retrenchment' that will emerge, a cultural turn that denies our data to others. Enterprising inventors could launch new apps and new businesses that would shield user data from the prying eyes of algorithms. Can the secret life of data continue to hum along if users make the choice not to participate?

The original Luddites smashed power looms and other technologies that they perceived were threatening their jobs and livelihoods. We typically draw two conclusions from the experiences of 19th century Luddites: (1) that Luddites attempt to halt technological 'progress,' and (2) Luddites always lose. What if 21st-century Luddites were to 'win?' Perhaps users do not need the permission of designers and technologists to make choices about which technologies to use and how to use them? What choices will users make that might be at odds with those of the Internet's designers? Even though technologists might develop them, perhaps users will choose not to wear networked glasses or contact lenses; perhaps they will collectively proclaim 'I do not want to be implanted!' Marxists like to talk about 'exhaustion:' the exhaustion of late capitalism or, for some literary critics, the exhaustion of print. Exhaustion here means having run its course, having used up all of its energy.[21] Will we see a widespread cultural exhaustion with Internet connectivity?

Extending cognition out beyond the brain, coupling with the cognitive objects the brain has conceived, is an ancient and human impulse. Brain–Internet coupling is not evidence of humanity becoming stupid or a sign that we are losing our humanness. Indeed, I believe it to be an expression of our humanity. What is uncertain at this stage is what that

DOI: 10.1057/9781137460950.0008

coupling will look like, what degree of intimacy we will reach with this latest extension of our symbolic storage system. If we do end up implanting devices such that we interact directly and intimately with the Internet, or that we interface with information arriving via all of our senses, it will be because we will have managed to transgress a number of limits and frontiers. What will be the result when our impulse to extend cognition confronts any number of boundaries and limits upon that impulse?

Notes

1　On scenario development and scenario thinking, see Peter Schwartz, *The Art of the Long View: Paths to Strategic Insight for Yourself and Your Company* (New York: Currency Doubleday, 1996); and Kees van der Heijden, *Scenarios: The Art of Strategic Conversation* (New York: John Wiley and Sons, 1996).

2　Michio Kaku, *Physics of the Future: How Science Will Shape Human Destiny and Our Daily Lives by the Year 2100* (New York: Doubleday, 2011), 39–41.

3　Kaku, *Physics of the Future*, 189.

4　Ibid., 189–195.

5　Jeff Stibel, *Breakpoint* (New York: Palgrave Macmillan, 2013).

6　David Auerbach, 'Yes, Your Internet Is Getting Slower,' *Slate*, May 14, 2014, http://www.slate.com/articles/technology/technology/2014/05/network_neutrality_dinosaurs_like_time_warner_and_at_t_have_nothing_to_worry.html; Marvin Ammori, 'We're About to Lose Net Neutrality – And the Internet as We Know It,' *Wired* November 14, 2013 http://www.wired.com/2013/11/so-the-internets-about-to-lose-its-net-neutrality/

7　'Britain's energy crisis: How long till the lights go out?' *The Economist* (August 6, 2009), http://www.economist.com/node/14167834; Federal Reserve Bank of St. Louis , 'Electricity: The Next Looming Energy Crisis?' (October 3, 2006), http://www.stlouisfed.org/newsroom/displayNews.cfm?article=312; Stephen M. Millett, 'Trends Impacting CPA Customers and Services to 2025,' Presentation to the Ohio Society of CPAs, February 26, 2009, http://www.ohioscpa.com/docs/conference-outlines/25_macro-trends-impacting-your-customers-and-services-to-2025.pdf?sfvrsn=2

8　'A Super Solar Flare,' *NASA Science News*, 2008, http://science1.nasa.gov/science-news/science-at-nasa/2008/06may_carringtonflare/ Glenn Harlan Reynolds, 'Solar Flare poses huge threat,' *USA Today,* June 28, 2013, http://www.usatoday.com/story/opinion/2013/06/26/solar-flare-electrical-threat-column/2461313/ 'As electronic technologies have become more sophisticated and more embedded into everyday life, they have also become more vulnerable to solar activity. On Earth, power lines and long-distance

DOI: 10.1057/9781137460950.0008

telephone cables might be affected by auroral currents, as happened in 1989 [when a similar solar flare disrupted electric power transmission from the Hydro Québec generating station]. Radar, cell phone communications, and GPS receivers could be disrupted by solar radio noise. Experts who have studied the question say there is little to be done to protect satellites from a Carrington-class flare. In fact, a recent paper estimates potential damage to the 900-plus satellites currently in orbit could cost between $30 billion and $70 billion.'

9   Wolfgang Ernst, *Digital Memory and the Archive* (Minneapolis: University of Minnesota Press, 2012), 85–86. 'The archival infrastructure in the case of the Internet is only ever temporary, in response to its permanent dynamic rewriting. Ultimate knowledge (the old encyclopedia model) gives way to the principle of permanent rewriting or addition (Wikipedia). The memory spaces geared to eternity are replaced by series of temporally limited entries with internal expiry dates that are as reconfigurable as the rhetorical mechanisms of the ars memoriae once were … Ostensibly the largest digital archive, the Internet is in fact a collection or assembly … The archive is defined as a given, preselected quantity of documents evaluated according to their worth for being handed down. The Internet, on the other hand, is an aggregate of unpredictable texts, sounds, images, data, and programs.'

10 Institute for Digital Research in the Humanities, University of Kansas, 'Return to the Material Conference,' September 14, 2013, http://idrh.ku.edu/dhforum2013/return-to-the-material

11 Claudius Gros, Gregor Kaczor and Dimitrije Markovic, 'Neuropsychological constraints to human data production on a global scale,' *European Physical Journal B*, 85: 28, 2012, http://arxiv.org/pdf/1111.6849v1.pdf See also 'Human Brain Is Limiting Global Data Growth, Say Computer Scientists,' *MIT Technology Review* (December 1, 2011), http://www.technologyreview.com/view/426246/human-brain-is-limiting-global-data-growth-say-computer-scientists/

12 Dave Eggers, *The Circle* (McSweeney's Books, 2013), 190.

13 Alexis Madrigal, 'Bolt Is Freaky Fast, But Nowhere Near Human Limits,' *Wired* August 25, 2008, http://www.wired.com/2008/08/bolt-is-freaky/; Ed Yong, 'Will we ever … run 100m in under nine seconds,' *BBC Future*, July 13, 2012, http://www.bbc.com/future/story/20120712-will-we-ever-run-100m-in-9-secs

14 Narrative Science, http://narrativescience.com/

15 Peter Swirski, *From Literature to Biterature: Lem, Turing, Darwin, and Explorations in Computer Literature, Philosophy of Mind, and Cultural Evolution* (McGill-Queen's University Press, 2013), 197.

16 Jaron Lanier, *You Are Not A Gadget: A Manifesto* (New York: Alfred A. Knopf, 2010), 5.

DOI: 10.1057/9781137460950.0008

17   Lanier, *You Are Not A Gadget*, 5–6. 'Technologists don't use persuasion to influence you – or, at least, we don't do it very well. There are a few master communicators among us (like Steve Jobs), but for the most part we aren't particularly seductive.

We make up extensions to your being, like remote eyes and ears (webcams and mobile phones) and expanded memory (the world of details you can search for online). These become the structures by which you connect to the world and other people. These structures in turn can change how you conceive of yourself and the world. We tinker with your philosophy by direct manipulation of your cognitive experience, not indirectly through argument. It takes only a tiny group of engineers to create technology that can shape the entire future of human experience with incredible speed. Therefore, crucial arguments about the human relationship with technology should take place between developers and users before such direct manipulations are designed.'

18   See Vance Packard, *The Waste Makers* (New York: D. McKay Co., 1960), 78–91; Victor Papanek, *Design for the Real World: Human Ecology and Social Change* (Toronto: Bantam Books, 1973).

19   Brian Winston, *Media, Technology and Society: A History: From the Telegraph to the Internet* (London: Routledge, 1998), 5–7.

20   BBC News, 'EU court backs "right to be forgotten" in Google case,' May 13, 2014, http://www.bbc.com/news/world-europe-27388289; Dave Lee, 'What is the "right to be forgotten?"', BBC News, May 13, 2014, http://www.bbc.com/news/technology-27394751; *The Economist*, 'On being forgotten,' http://www.economist.com/news/leaders/21602219-right-be-forgotten-sounds-attractive-it-creates-more-problems-it-solves-being

21   'Towards a Cultural History of Exhaustion,' *Centre for Medical Humanities Blog, Durham University* (May 8, 2013), http://medicalhumanities.wordpress.com/2013/05/08/towards-a-cultural-history-of-exhaustion/; John Barth, 'The Literature of Exhaustion,' in *The Friday Book: Essays and Other Non-Fiction* (Baltimore: The John Hopkins University Press, 1997), 62–76.

Conclude

Abstract: *The goal of this essay was to address concerns that the Internet is harming our brains by placing our current historical moment in a deeper historical context. By expanding our temporal context we may make better sense of the present moment, and in the process draw different lessons than the-Internet-is-damaging-our-brains narrative. This essay also explored potential future paths that our coupled cognition might take. Considering different scenarios allows us to identify those drivers that will determine the future shape of the system called the brain–Internet interface. The most important driver is the evolutionary impulse to expand our cognitive capacity through symbolic objects fashioned by the human mind. The Internet is a coupled system with our brain that will help define what it means to be human.*

Keywords: future; Internet

Staley, David J. *Brain, Mind and Internet: A Deep History and Future*. Basingstoke: Palgrave Macmillan, 2014. DOI: 10.1057/9781137460950.0009.

DOI: 10.1057/9781137460950.0009

My goal in this essay was two-fold. First, I wished to directly address concerns that the Internet is harming our brains by placing our current historical moment in a much longer historical context. Many commentators contextualize the issue too narrowly, and place the Internet within the temporal orbit of the culture of The Book. To view the Internet as harmful is to identify the many ways that it is not like The Book; the problem with the Internet is that it encourages us to think in a manner different from that encouraged by books. Thus, the Internet does not facilitate the sustained engagement with a text, therefore it is harmful, or the Book encourages linear thinking, and the Internet is associative, therefore the Internet is damaging our brains. The Book is but only one cultural object, one form of external symbolic storage, with which our brains have coupled. We should expand our temporal horizon beyond The Book in order to grasp the meaning of the current brain–Internet coupling. By expanding our temporal context (backward and forward) I hope to make sense of the present moment, and in the process allow us to draw different lessons about the present, different lessons than the-Internet-is-damaging-our-brains narrative.

Second, given that deep historical contextualization, I wanted to explore some potential future paths our coupled cognition might take. What we term 'the future' is the state of some system at a point in the future. That system could transform in any number of ways, and thus I identified three states or paths that system might travel: query, interface, and limit. If futurists caution that we cannot predict the future with any certainty, that the system under consideration is so intrinsically complex as to defy simple trend extrapolation, then what is the point of the speculation undertaken in this essay? Why make all of these stories and scenarios? Like history, futuring is a meaning-making activity. The historian William McNeill once wrote that 'Myth and history are close kin inasmuch as both explain how things got to be the way they are by telling some sort of story.'[1] I have long believed that one of the main functions of the historian is to be a sense-maker, not only with regard to the events of the past but to our contemporary moment. Historians make sense of things because we understand how the meaning of events is influenced by the surrounding context.

Futuring is also a kind of mythmaking. Like McNeill's definition above, myth here does not mean falsehood or make-believe, but a narrative that purports to understand the meaning of events, a story that allows us to make sense. When faced with the formless void that is the future, any

DOI: 10.1057/9781137460950.0009

narrative is better than no narrative, because that narrative helps us to orient our thoughts, guide our actions (or our resistance), and helps us to see the implications of the choices we are making at the present.[2]

Because the future is without form, the futurist serves as a 'form giver.' (The original German term for design was Gestaltung: 'to give form.') Scenarios are one way to give form to the future, to fill that formless void with something that stands in for the future. The philosopher Frank Ankersmit observes that historical narratives are really 'proposals' about the past (not the actual past itself). Because the past no longer exists, we require something to 'stand in' for the past, which is the role usually played by a historical narrative.[3] Similarly, a scenario is a 'proposal' about the future in that it 'stands in' for that which does not yet exist, aiding us to make choices and decisions today. My 'Query,' 'Interface' and 'Limit' scenarios are narratives that 'stand in' for the future, with the goal of providing meaning and a sense of direction about the potential course for the brain–Internet interface.

Considering different scenarios allows us to identify those factors, or drivers, which will determine the future shape of the system called brain–Internet interface. Those drivers include:

▸ the desires of the designers, technologists, entrepreneurs, and corporations who will develop the next generation of cognitive technologies;
▸ the needs and aspirations of users of those cognitive technologies, and how their use of these tools will shape the future of cognition, the future architecture of the mind;
▸ our technological capabilities and the physical and conceptual limits on those technologies;
▸ the means by which we will adapt to our current environment, and the ways we will alter that environment through our cognitive technologies;
▸ the limits imposed by our brains/bodies on the development and use of the Internet as a cognitive prosthetic.

Perhaps the most important driver, however, is the evolutionary impulse to expand our cognitive capacity through symbolic objects fashioned by the human mind. Humans create an external extension to the brain just as surely as a spider spins a web. The Internet is nothing more or less than the next stage in the exfoliation of external cognition. It is a coupled system with our brain which will help define what it means to be human.

DOI: 10.1057/9781137460950.0009

I once had a member of my staff who was a very bright, very capable young programmer. I was in awe of his range of talents and abilities, which far exceed my own. (He now works for Cisco.) One day, I was looking over his shoulder as he was working on a coding problem I had laid out before him. He was surfing the Internet (and not working?). When I asked what he was doing, he said he was looking to copy several lines of code from someone in a user group who might know the answer. It seems that my programmer did not hold all of that knowledge in his head. When he does not know how to do something, he will Google a query, then choose from among the best-looking responses (a judgment he has honed over the years) to locate someone's post or blog who has answered the question and has provided the code he needs. 'This is how most coders work,' my assistant informed me. 'When we don't know how to do something, we look it up on the Internet.'

Since I began this essay with a cartoon, perhaps it is fitting that I conclude with one as well. Although it is not as widely syndicated as Doonesbury, *xkcd* also captures larger technologically influenced cultural trends. The particular cartoon I have in mind is an open letter to 'non-computer people.'4 The anonymous author states '["Computer people"] don't magically know how to do everything in every program. When we help you, we're usually just doing this:' The cartoon is a flow chart that concludes with an instruction that reads 'Google the name of the program plus a few words related to what you want to do. Follow any instructions.' Reflect on the truth of this cartoon: 'computer people,' like my student assistant, frequently rely on the Cloud for their information and knowledge, knowledge they access just-in-time to solve a problem at hand.

In my subsequent conversations with my student assistant, it is clear that he took a similar just-in-time attitude toward other academic subjects. After taking an in-class exam where he was asked to recount a narrative and to write out short-answer identifications of important events, this student asked me with some exasperation, 'Why should I have to remember all of this stuff when it is just as easy for me to look it up on the Internet over my smartphone.' To reiterate, I considered him to be a very good student and an exceptionally intelligent young man. I would never describe him as lazy or intellectually slothful, like the Doonesbury character I described in the preface. Yet it is clear that he takes a very different attitude toward the acquisition and use of information and knowledge. For my young assistant, intelligence is not a matter of 'know what' but of 'know how:' he knows how to seek out the right information that he has adjudged to be useful and applies it to solve the problem at hand. This, to me, is the very definition of the 'just-in-time' ethic that defines our current Internet moment, an ethic, it turns out, that is the result of a very long history.

Notes

1   William McNeill, 'Mythistory, or Truth, Myth, History and Historians,' in *Mythistory and Other Essays* (Chicago: University of Chicago Press, 1986), 4.
2   Kees van der Heijden, *Scenarios: The Art of Strategic Conversation* (New York: John Wiley and Sons, 1996), 36.
3   F.R. Ankersmit, *Historical Representation* (Stanford: Stanford University Press, 2001), 89.
4   http://imgs.xkcd.com/comics/tech_support_cheat_sheet.png

Bibliography

Ammori, Marvin. 'We're About to Lose Net Neutrality – And the Internet as We Know It,' *Wired*, November 14, 2013 http://www.wired.com/2013/11/so-the-internets-about-to- lose-its-net-neutrality/

Ankersmit, F.R. *Historical Representation* (Stanford: Stanford University Press, 2001).

Armstrong, David F., William C. Stokoe, and Sherman E. Wilcox. *Gesture and the Nature of Language* (Cambridge: Cambridge University Press, 1995).

Auerbach, David. 'Yes, Your Internet Is Getting Slower,' *Slate*, May 14, 2014, http://www.slate.com/articles/technology/technology/2014/05/network_neutrality_dinosa urs_like_time_warner_and_at_t_have_nothing_to_worry.html

Barabasi, Albert-Laszlo. *Bursts: The Hidden Pattern Behind Everything We Do* (New York: Dutton, 2010).

Barth, John. 'The Literature of Exhaustion,' in *The Friday Book: Essays and Other Non-Fiction.* (Baltimore: The John Hopkins University Press, 1997), 62–76.

BBC News, 'EU court backs "right to be forgotten" in Google case,' May 13, 2014, http://www.bbc.com/news/world-europe-27388289

Bolter, J. David. *Writing Space: Computers, Hypertext, and the Remediation of Print*, second edition (Mahwah, N.J.: Lawrence Erlbaum Associates, 2001).

Braudel, Fernand. trans. by Sarah Matthews, *On History* (Chicago: The University of Chicago Press, 1980).

Bredekamp, Horst. *The Lure of Antiquity and the Cult of the Machine: The Kunstkammer and the Evolution of*

Nature, Art and Technology (Princeton: Markus Wiener Publishers, 1995).

Brockman, John. *Is the Internet Changing the Way You Think? The Net's Impact on our Minds and Future* (New York: Harper Perennial, 2011).

Brumfield, C. Russell. *Whiff! The Revolution of Scent Communication in the Information Age* (New York: Quimby Press, 2008).

Bush, Vannevar. 'As We May Think,' *The Atlantic*, July 1, 1945, http://www.theatlantic.com/magazine/archive/1945/07/as-we-may-think/303881/

Carr, Nicholas. 'The Library of Utopia,' *MIT Technology Review*, April 25, 2012, http://www.technologyreview.com/featuredstory/427628/the-library-of-utopia/

Carr, Nicholas. *The Shallows: What the Internet is Doing to our Brains* (New York: W.W. Norton, 2010).

Carr, Nicholas. 'Is Google Making Us Stupid?' *The Atlantic*, July/August 2008, http://www.theatlantic.com/magazine/archive/2008/07/is-google-making-us-stupid/6868/

Centre for Medical Humanities Blog, Durham University. 'Towards a Cultural History of Exhaustion,' May 8, 2013, http://medicalhumanities.wordpress.com/2013/05/08/towards-a-cultural-history-of- exhaustion/

Chorost, Michael. *World Wide Mind: The Coming Integration of Humanity, Machines and the Internet* (New York: Free Press, 2011).

Clark, Andy. *Supersizing the Mind: Embodiment, Action, and Cognitive Extension* (New York: Oxford University Press, 2008).

Clark, Andy. *Natural-Born Cyborgs: Minds, Technologies, and the Future of Human Intelligence* (New York: Oxford University Press, 2003).

Clark, Andy and David J. Chalmers. 'The Extended Mind,' in Richard Menary, ed. *The Extended Mind* (Cambridge, MA: MIT Press, 2010), 27.

Dehaene, Stanislas. *Reading in the Brain: The Science and Evolution of a Human Invention* (New York: Viking, 2009).

Donald, Merlin. *A Mind So Rare: The Evolution of Human Consciousness* (New York: W.W. Norton, 2001).

Donald, Merlin. *Origins of the Modern Mind: Three Stages in the Evolution of Culture and Cognition* (Cambridge: Harvard University Press, 1991).

Egan, Kieran. *The Future of Education: Reimagining our Schools from the Ground Up* (New Haven: Yale University Press, 2008).

Ernst, Wolfgang. *Digital Memory and the Archive* (Minneapolis: University of Minnesota Press, 2012).

DOI: 10.1057/9781137460950.0010

Federal Reserve Bank of St. Louis , 'Electricity: The Next Looming Energy Crisis?' October 3, 2006, http://www.stlouisfed.org/newsroom/displayNews.cfm?article=312

Foreman, Richard. 'The Pancake People, or, "The Gods are Pounding my Head,"' *Edge, The Third Culture*, http://www.edge.org/3rd_culture/foreman05/foreman05_index.html

Fox, Kate. *The Smell Report: An overview of facts and findings*, Social Issues Research Centre, http://www.sirc.org/publik/smell.pdf

Goertzel, Ben. *Creating Internet Intelligence: Wild Computing, Distributed Digital Consciousness, and the Emerging Global Brain* (New York: Kluwer Academic/Plenum Publishers, 2002).

Gore, Al. *The Future: Six Drivers of Global Change* (New York: Random House, 2013).

Gros, Claudius, Gregor Kaczor, and Dimitrije Markovic. 'Neuropsychological constraints to human data production on a global scale,' *European Physical Journal B*, 85: 28 (2012), http://arxiv.org/pdf/1111.6849v1.pdf

Halpin, Harry, Andy, Clark, and Michael, Wheeler. 'Towards a Philosophy of the Web: Representation, Enaction, Collective Intelligence,' Proceedings of the Web Science Conference: Extending the Frontiers of Society On-Line , April 26–27, 2010, http://citeseerx.ist.psu.edu/viewdoc/download;jsessionid=DCBD0BC4BD08A6E2E4DF3 A2EDCFBC023?doi=10.1.1.415.282&rep=rep1&type=pdf

Harpham, Geoffrey Galt. *The Humanities and the Dream of America* (Chicago: University of Chicago Press, 2011), 23.

Hayles, N. Katherine. *How We Think: Digital Media and Contemporary Technogenesis* (Chicago: University of Chicago Press, 2012).

Hillis, Ken, Michael, Petit, and Kylie, Jarrett. *Google and the Culture of Search* (New York: Routledge, 2013).

Hobart, Michael E. and Zachary S. Schiffman. *Information Ages: Literacy, Numeracy, and the Computer Revolution* (Baltimore: Johns Hopkins University Press, 1998).

'Internet Human | Human Internet Map,' *Institute for the Future*, May 2013, http://www.iftf.org/uploads/media/IFTF_TH12-InternetHuman_map_rdr.pdf

Johnson, Brian David. 'The Secret Life of Data in the Year 2020,' *The Futurist*, July–August 2012, 20–23.

Kahneman, Daniel. *Thinking, Fast and Slow* (New York: Farrar, Straus and Giroux, 2011).

DOI: 10.1057/9781137460950.0010

Kaku, Michio. *Physics of the Future: How Science Will Shape Human Destiny and Our Daily Lives by the Year 2100* (New York: Doubleday, 2011).

Kamenetz, Anya. 'How TED Became the New Harvard,' *Fast Company* 82, September 2010, http://www.fastcompany.com/1677383/how-ted-connects-idea-hungry-elite

Kirschenbaum, Matthew G. 'The Remaking of Reading: Data Mining and the Digital Humanities,' *NSF Symposium on Next Generation of Data Mining and Cyber-Enabled Discovery for Innovation*, October 11, 2007, http://www.csee.umbc.edu/~hillol/NGDM07/abstracts/talks/MKirschenbaum.pdf

Kurzweil, Ray. *How to Create a Mind: The Secret of Human Thought Revealed* (New York: Viking, 2012).

Kurzweil, Ray. *The Singularity Is Near: When Humans Transcend Biology* (New York: Penguin, 2005).

Lakoff, George and Mark, Johnson. *Philosophy in the Flesh: The Embodied Mind and Its Challenge to Western Thought* (New York: Basic Books, 1999).

Lanier, Jaron. *You Are Not A Gadget: A Manifesto* (New York: Alfred A. Knopf, 2010).

Lee, Dave. 'What is the "Right to be Forgotten?",' BBC News, May 13, 2014, http://www.bbc.com/news/technology-27394751

Licklider, J.C.R. 'Man-Computer Symbiosis,' *IRE Transactions on Human Factors in Electronics*, HFE-1 (March 1960), 4–11.

Lindstrom, Martin. *Brand Sense: Build Powerful Brands Through Touch, Taste, Smell, Sight and Sound* (New York: Free Press, 2005).

Madrigal, Alexis. 'Bolt Is Freaky Fast, But Nowhere Near Human Limits,' *Wired* August 25, 2008, http://www.wired.com/2008/08/bolt-is-freaky/

Malafouris, Lambros. *How Things Shape the Mind: A Theory of Material Engagement* (Cambridge: MIT Press, 2013).

May-raz, Eran and Daniel Laz. *Sight*, http://vimeo.com/46304267

McCullough, Malcolm. *Ambient Commons: Attention in the Age of Embodied Information* (Cambridge: MIT Press, 2013).

McNeill, David. *Hand and Mind: What Gestures Reveal About Thought* (Chicago: University of Chicago Press, 1992).

McNeill, William. *Mythistory and Other Essays* (Chicago: University of Chicago Press, 1986).

Millett, Stephen M. 'Trends Impacting CPA Customers and Services to 2025,' Presentation to the Ohio Society of CPAs, February 26, 2009,

DOI: 10.1057/9781137460950.0010

http://www.ohioscpa.com/docs/conference-outlines/25_macro-trends-impacting-your-customers-and-services-to-2025.pdf?sfvrsn=2

MIT Technology Review, 'Human Brain Is Limiting Global Data Growth, Say Computer Scientists,' December 1, 2011, http://www.technologyreview.com/view/426246/human- brain-is-limiting-global-data-growth-say-computer-scientists/

NASA Science News. 'A Super Solar Flare,' 2008, http://science1.nasa.gov/science- news/science-at-nasa/2008/06may_carringtonflare/

Nielsen, Michael. *Reinventing Discovery: The New Era of Networked Science* (Princeton: Princeton University Press, 2012).

Norman, Donald. *The Design of Future Things* (New York: Basic Books, 2007).

Odlyzko, Andrew. 'The myth of Internet time,' *MIT Technology Review*, April 1, 2001, http://www.technologyreview.com/review/400952/the-myth-of-internet-time/

Packard, Vance. *The Waste Makers* (New York: D. McKay Co., 1960).

Palmer, Jason. 'Brain works more like internet than "top down" company,' *BBC News*, 10 August 2010, http://www.bbc.co.uk/news/science-environment-10925841

Papanek, Victor. *Design for the Real World: Human Ecology and Social Change* (Toronto: Bantam Books, 1973).

Pasztory, Esther. *Thinking With Things: Toward a New Vision of Art* (Austin: University of Texas Press, 2005).

Pink, Daniel. *A Whole New Mind: Moving from the Information Age to the Conceptual Age* (New York: Riverhead Books, 2005).

Pinker, Steven. *The Language Instinct: How the Mind Creates Language* (New York: Harper Perennial, 1995).

Plato, *The Phaedrus*, http://www.units.muohio.edu/technologyandhumanities/plato.htm

Prensky, Marc. *Brain Gain: Technology and the Quest for Digital Wisdom* (New York: Palgrave Macmillan, 2012).

Renfrew, Colin. *Prehistory: The Making of the Human Mind* (New York: The Modern Library, 2007).

Reynolds, Glenn Harlan. 'Solar Flare poses huge threat,' *USA Today*, June 28, 2013, http://www.usatoday.com/story/opinion/2013/06/26/solar-flare-electrical-threat- column/2461313/

Rickert, Thomas. *Ambient Rhetoric: The Attunements of Rhetorical Being* (Pittsburgh: University of Pittsburgh Press, 2013).

DOI: 10.1057/9781137460950.0010

Rowlands, Mark. *The New Science of the Mind: From Extended Mind to Embodied Phenomenology* (Cambridge, MA: MIT Press, 2010).

Rupert, Robert D. *Cognitive Systems and the Extended Mind* (New York: Oxford University Press, 2009).

Sapolsky, Robert. 'People who can intuit in six dimensions,' in John Brockman, ed. *This Will Change Everything: Ideas That Will Shape the Future* (New York: Harper Perennial, 2010), 366–369.

Schwartz, Peter. *The Art of the Long View: Paths to Strategic Insight for Yourself and Your Company* (New York: Currency Doubleday, 1996).

Selingo, Jeffrey J. *College (Un)Bound: The Future of Higher Education and What It Means for Students* (Boston: New Harvest, 2013).

Singer, Peter 'Whither the dream of the universal library?' *The Guardian* (April 19, 2011), http://www.theguardian.com/commentisfree/2011/apr/19/moral-imperative-create- universal-library

Smail, Daniel Lord. *On Deep History and the Brain* (Berkeley: University of California Press, 2008).

Stafford, Barbara Maria. *Visual Analogy: Consciousness as the Art of Connecting* (Cambridge, MA: MIT Press, 1999).

Stibel, Jeff. *Breakpoint* (New York: Palgrave Macmillan, 2013).

Stibel, Jeff. *Wired for Thought: How the Brain is Shaping the Future of the Internet* (Boston: Harvard Business Press, 2009).

Thackera, John. *In the Bubble: Designing in a Complex World* (Cambridge: MIT Press, 2005).

Thaler, Richard H. and Cass R. Sunstein. *Nudge: Improving Decisions About Health, Wealth, and Happiness* (New Haven: Yale University Press, 2008).

The Economist. 'Britain's energy crisis: How long till the lights go out?' August 6, 2009, http://www.economist.com/node/14167834

The Economist. 'On being forgotten,' http://www.economist.com/news/leaders/21602219-right- be-forgotten-sounds-attractive-it-creates-more-problems-it-solves-being

Thompson, Clive. 'A Sense of Place,' *Wired* (February 2013), 34.

Topol, Eric. *The Creative Destruction of Medicine: How the Digital Revolution Will Create Better Health Care* (New York: Basic Books, 2012).

van der Heijden, Kees. *Scenarios: The Art of Strategic Conversation* (New York: John Wiley and Sons, 1996).

Wagner, Caroline S. *The New Invisible College: Science for Development* (Washington, D.C.: Brookings Institution Press, 2008).

DOI: 10.1057/9781137460950.0010

Wattenburg, Martin. 'Map of the Market' (1998) http://www.bewitched.com/marketmap.html

Wells, H.G. *World Brain* (Garden City, New York: Doubleday, Doran and Co., 1938).

Winston, Brian. *Media, Technology and Society: A History: From the Telegraph to the Internet* (London: Routledge, 1998).

Wisneski, Craig, Hiroshi Ishii, Andrew Dahley, Matt Gorbet, Scott Brave, Brygg Ullmer, and Paul Yarin. 'Ambient Displays: Turning Architectural Space into an Interface between People and Digital Information,' *Proceedings of the First International Workshop on Cooperative Buildings* (CoBuild '98), February 25–26 (1998), 22–32.

Yong, Ed. 'Will we ever … run 100m in under nine seconds,' *BBC Future*, July 13, 2012, http://www.bbc.com/future/story/20120712-will-we-ever-run-100m-in-9-secs

DOI: 10.1057/9781137460950.0010

Index

DOI: 10.1057/9781137460950.0011

DOI: 10.1057/9781137460950.0011